AF885269

Autodesk 3ds Max 2017 for Novices

(Learn By Doing)

CADSoft Technologies

Autodesk 3ds Max 2017 for Novices: Learn By Doing

CADSoft Technologies

CADSoft Technologies

ISBN 978-1-64057-029-0

NOTICE TO THE READER

Publisher does not warrant or guarantee any of the products described in the text or perform any independent analysis in connection with any of the product information contained in the text. Publisher does not assume, and expressly disclaims, any obligation to obtain and include information other than that provided to it by the manufacturer.

The reader is expressly warned to consider and adopt all safety precautions that might be indicated by the activities herein and to avoid all potential hazards. By following the instructions contained herein, the reader willingly assumes all risks in connection with such instructions.

The publisher makes no representation or warranties of any kind, including but not limited to, the warranties of fitness for particular purpose or merchantability, nor are any such representations implied with respect to the material set forth herein, and the publisher takes no responsibility with respect to such material. The publisher shall not be liable for any special, consequential, or exemplary damages resulting, in whole or part, from the reader's use of, or reliance upon this material.

DEDICATION

To students, who are dedicated to learning new technologies

THANKS

To the faculty and employees of CADSoft Technologies
for their cooperation

Online Training Program Offered by CADSoft Technologies

CADSoft Technologies provides effective and affordable virtual online training on various software packages including Computer Aided Design, Manufacturing and Engineering (CAD/CAM/CAE), computer programming languages, animation, architecture, and GIS. The training is delivered 'live' via Internet at any time, any place, and at any pace to individuals as well as the students of colleges, universities, and CAD/CAM/CAE training centers. The main features of this program are:

Training for Students and Companies in a Classroom Setting

Highly experienced instructors and qualified Engineers at CADSoft Technologies conduct the classes under the guidance of Prof. Sham Tickoo of Purdue University Northwest, USA. This team has authored several textbooks that are rated "one of the best" in their categories and are used in various colleges, universities, and training centers in North America, Europe, and in other parts of the world.

Training for Individuals

CADSoft Technologies with its cost effective and time saving initiative strives to deliver the training in the comfort of your home or work place, thereby relieving you from the hassles of traveling to training centers.

Training Offered on Software Packages

CADSoft Technologies provide basic and advanced training on the following software packages:

CAD/CAM/CAE: *CATIA, Pro/ENGINEER Wildfire, SOLIDWORKS, Autodesk Inventor, Solid Edge, NX, AutoCAD, AutoCAD LT, Customizing AutoCAD, AutoCAD Electrical, EdgeCAM, and ANSYS*

Architecture and GIS: *Autodesk Revit (Architecture/Structure/MEP), Autodesk Navisworks, ETABS, Bentley STAAD.Pro, AutoCAD Raster Design, ArcGIS, AutoCAD Civil 3D, AutoCAD Map 3D, Oracle Primavera P6, MS Project*

Animation and Styling: *Autodesk 3ds Max, 3ds Max Design, Autodesk Maya, Blender, Pixologic ZBrush, and CINEMA 4D*

Computer Programming: *C++, VB.NET, Oracle, AJAX, PHP/MYSQL, and Java*

Personality Development: *Personality Development and Engineering Ethics/Soft Skills Course*

Table of Contents

Chapter 5: Splines and Extended Splines

Chapter 6: Modifying Splines

Chapter 7: Materials and Maps

Chapter 8: Modifying 3D Mesh Objects

Chapter 9: Graphite Modeling Technique

Chapter 10: Compound Objects

Chapter 11: Modifiers

Chapter 12: Lights and Cameras

This page is intentionally left blank

Preface

Autodesk 3ds Max 2017

Autodesk 3ds Max, developed by Autodesk Inc., provides powerful tools for 3D modeling, animation, rendering, dynamics, and compositing. This enables game developers, visual effects artists, architects, designers, engineers, and visualization specialists to create stunning artwork. Additionally, the intuitive user interface and workflow tools of 3ds Max 2017 have made the job of design visualization specialists easier.

Autodesk 3ds Max 2017 for Novices: Learn by Doing comprises of solved tutorials as well as exercises to help user to learn and practice 3ds Max for detailed description of concepts and tools.

Some of the main features of the book are as follows:

- **Real-world Architectural Projects as Tutorials**
  The author has used the real-world architectural projects as tutorials in this textbook so that the readers can correlate the tutorials with the real-time scenes.

- **Tips and Notes**
  Additional information related to various topics is provided to the users in the form of tips and notes.

Symbols Used in the Textbook

Note
The author has provided additional information to the users about the topic being discussed in the form of notes.

Tip
Special information and techniques are provided in the form of tips that helps in increasing the efficiency of the users.

Formatting Conventions Used in the Textbook

Please refer to the following list for the formatting conventions used in this textbook.

- Names of tools, buttons, options, renderer, rollouts, and tabs are written in boldface.

  Example: The **Select and Move** tool, the **Render** button, the **Modify** tab, **NVIDIA mental ray** renderer, the **Common Parameters** rollout, and so on.

- Names of dialog boxes, drop-downs, drop-down lists, spinners, areas, edit boxes, check boxes, and radio buttons are written in boldface.

  Example: The **Render Setup** dialog box, the **Look in** drop-down, the **Length** spinner, the **Bend** area, the **Angle** edit box, **Real-World Map Size** check box, the **Cube** radio button, and so on.

- Values entered in spinners are written in boldface.

  Example: Enter **10** in the **Length** spinner.

- Names of the files are italicized.

  Example: *c02_tut1.max*

Naming Conventions Used in the Textbook

Tool

If you click on an item in a panel of the ribbon and a command is invoked to create/edit an object or perform some action, then that item is termed as **tool**.

For example:
Select and Move tool, **Select and Link** tool, **Angle Snap Toggle** tool
Render Setup tool, **Select and Rotate** tool, **Align** tool

If you click on an item in a panel of the ribbon and a dialog box is invoked wherein you can set the properties to create/edit an object, then that item is also termed as **tool**.

For example:
Material Editor tool, **Render Setup** tool

Flyout

A flyout is an icon-based menu that contains tools with similar type of functions. Figure 1 shows the **Snap Toggle** flyout. The buttons having a small triangle at their lower right corner contain a flyout. Press and hold such a button; a flyout will be displayed, refer to Figure 1.

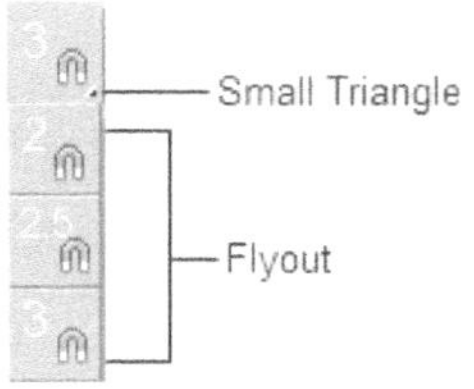

Figure 1 *The* ***Snap Toggle*** *flyout*

Right-click Menus

In Autodesk 3ds Max, the right-click menus provide quick access to the commonly used commands that are related to the current selection of an object. When you right-click on an object, a quad menu is displayed, as shown in Figure 2. Some of the options in the quad menu have an arrow on their right side. If you move the mouse pointer on these options, a cascading menu will be displayed, refer to Figure 2. But, if you right-click in the viewport, a shortcut menu will be displayed, refer to Figure 3.

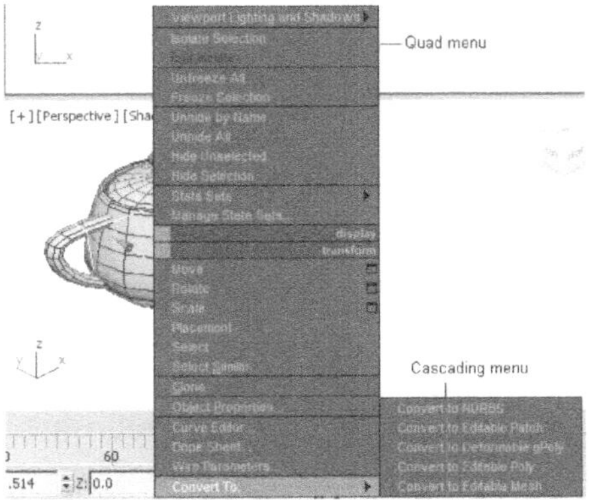

Figure 2 *The quad menu and the cascading menu*

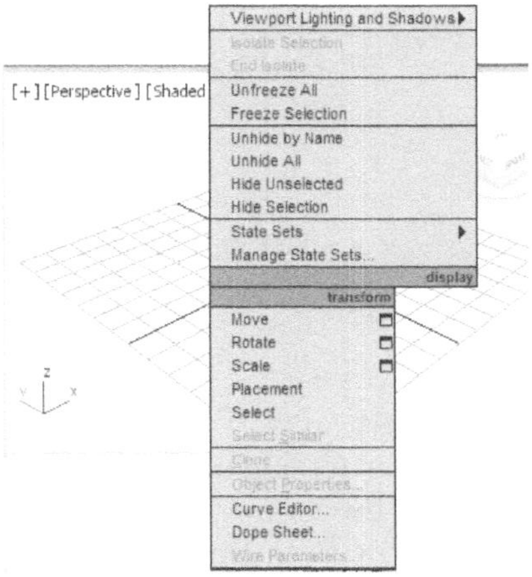

Figure 3 *The shortcut menu displayed on right-clicking in the viewport*

Button

The item in a dialog box that has a 3D shape is termed as **Button**. For example, **OK** button, **Cancel** button, **Render** button, and so on, refer to Figure 4.

Dialog Box

In this textbook, different terms are used for referring to the components of a dialog box. Refer to Figure 4 for the terminology used.

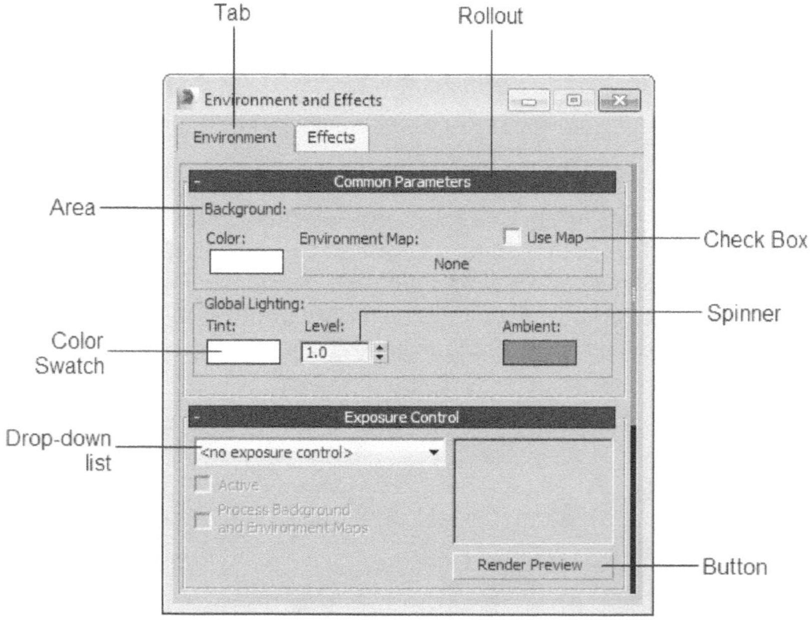

Figure 4 *Terminologies used for different components of a dialog box*

Drop-down List

A drop-down list is the one in which a set of options are grouped together. You can set various parameters using these options. You can identify a drop-down list with a down arrow on it. For example, **Reference Coordinate System** drop-down list, **Modifier List** drop-down list, and so on; refer to Figure 5.

Figure 5 *Selecting an option from the*
Modifier List *drop-down list*

Chapter 1

Introduction to Autodesk 3ds Max 2017

Learning Objectives

After completing this chapter, you will be able to:

- *Understand the Autodesk 3ds Max interface components*
- *Use controls for creating or modifying objects*

Autodesk 3ds Max INTERFACE COMPONENTS

The 3ds Max interface screen consists of different components, as shown in Figure 1-1.

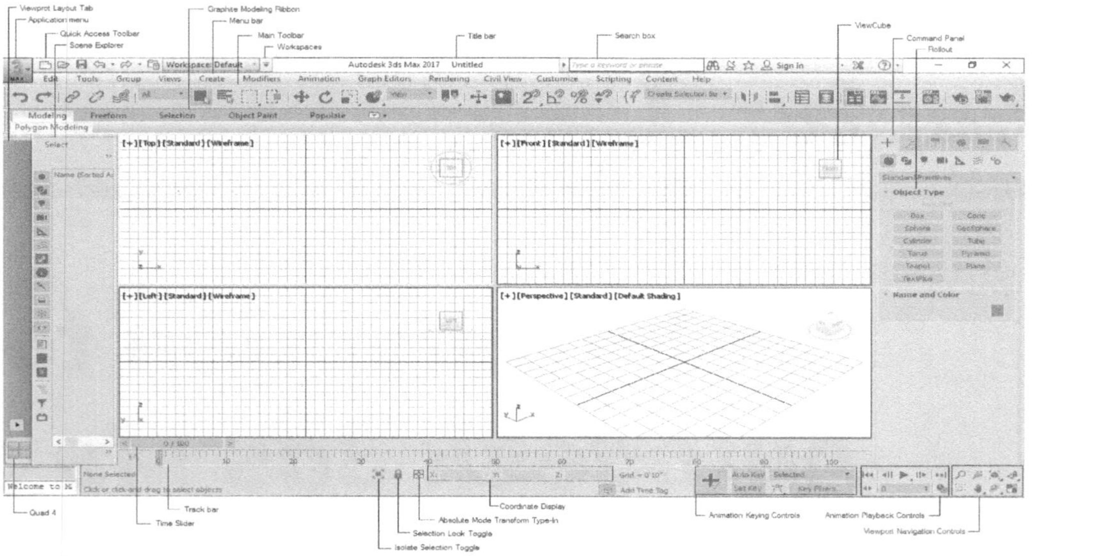

***Figure 1-1** Different screen components of Autodesk 3ds Max interface screen*

STARTING A NEW SCENE

Application Menu: New > New All
Quick Access Toolbar: New

You can open a new scene using the **New** tool from the **Quick Access Toolbar**. When you invoke the **New** tool, by default 3ds Max will display the **New Scene** message box, as shown in Figure 1-2. This dialog box displays a list of options to open a new scene.

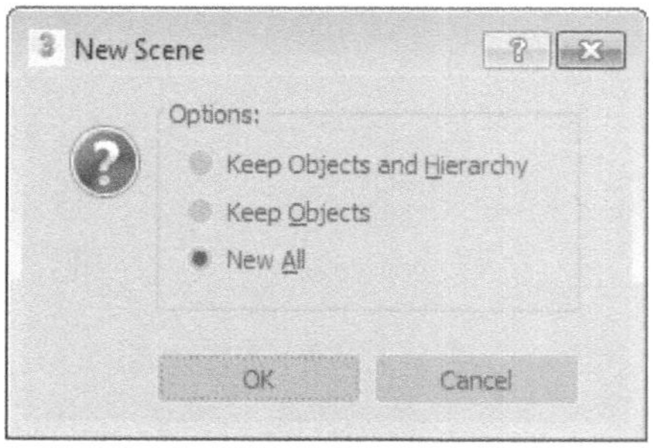

***Figure 1-2** The **New Scene** message box*

SAVING THE WORK

Application Menu: Save, Save As
Quick Access Toolbar: Save File

In 3ds Max, you need to save your work before you exit 3ds Max or turn off the system. Also, it is recommended that you save your files after regular time intervals, so that in the event of a power failure or an editing error, all work done by you is not lost and only the unsaved part is affected.

OPENING AN EXISTING SCENE

Application Menu: Open > Open
Quick Access Toolbar: Open File

You can open an existing scene that has been saved previously. There are two methods that can be used to open a scene: by using the **Application** menu or the **Quick Access Toolbar**.

QUITTING 3ds Max

You can exit 3ds Max by choosing **Exit 3ds Max** from the **Application Menu**. In case the scene has not been saved, the **Autodesk 3ds Max 2017** message box will be displayed. Note that if you choose **Don't Save** in this message box, all changes made in the current scene, till they were last saved, will be lost. You can also use the close button (**X**) of the main 3ds Max window (present in the title bar) to end the 3ds Max session.

This page is intentionally left blank

Chapter 2

Standard Primitives

Learning Objectives

After completing this chapter, you will be able to:
- *Understand selection techniques*
- *Create standard primitives*

INTRODUCTION

In this chapter you will learn the frequently used selection techniques in a max project. Also, with the help of a project you will learn to apply the various standard primitives available in 3dsMax.

The frrequently used selection techniques and the use of important standard primitives are discussed next.

SELECTION TECHNIQUES

In 3ds Max, you can select objects using various tools such as **Select Object**, **Select by Name**, **Select and Move**, and so on. These tools are discussed next.

Select Object Tool

Quad Menu:	Select
Main Toolbar:	Select Object
Keyboard:	Q

The **Select Object** tool is used to select one or more objects in the viewport. To select an object, choose this tool and move the cursor over the object; the cursor will convert into a selection cursor. Next, press the left mouse button; the object will be selected. To select more than one object at a time, hold the CTRL key and select the objects that you want to add to the selection. To remove an object from the selection, press and hold the ALT key and then click on the object that you want to remove from the selection.

Select and Move Tool

Quad Menu:	Move
Main Toolbar:	Select and Move
Keyboard:	W

The **Select and Move** tool is used to select and move the objects in the viewports. You can move an object by selecting it and then dragging the mouse along the X, Y, or Z axis. You can also move the selected object in the XY, YZ, or ZX plane. To move the selected object along one of the axes, choose this tool and then select the object; the move gizmo will be displayed. Move the cursor over the axis along which you want to move the object, press the left mouse button, and then drag the cursor. Similarly, to move the object in one of the planes, move the cursor over the plane displayed between the two axes; the plane will be highlighted in yellow. Next, press the left mouse button and drag the cursor.

Select and Rotate Tool

Quad Menu:	Rotate
Main Toolbar:	Select and Rotate
Keyboard:	E

The **Select and Rotate** tool is used to rotate the objects in the viewport along the X, Y, or Z axis. To rotate the object along one of the axes, choose the **Select and Rotate** tool, and then select the object; a rotate gizmo will be displayed along with the X, Y, and Z axes. Next, move the cursor over the axes along which you want to rotate the object, press the left mouse button, and then drag the cursor. When you rotate the object, a transparent slice will be displayed which will provide a visual representation of the direction and the degree of rotation, refer to Figure 2-1. Also, you can view the degree of rotation in the X, Y, and Z axes in the coordinates displayed in the Coordinate display area at the bottom of the screen.

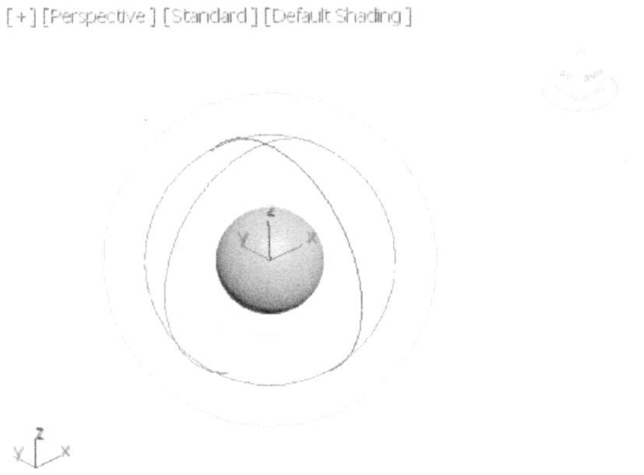

Figure 2-1 *The circular gizmo in the Perspective viewport displayed while rotating the object*

Select and Scale Tools

Quad Menu:	Scale
Main Toolbar:	Select and Scale
Keyboard:	R

There are three types of tools that are used to scale an object. These tools are available in the Select and Scale flyout and are discussed next.

Select and Uniform Scale

The **Select and Uniform Scale** tool is used to scale the objects proportionally along the three axes. To scale an object uniformly along all the three axes, choose the **Select and Uniform Scale** tool from the **Main Toolbar** and select the object; the scale gizmo will be displayed, as shown in Figure 2-2. Move the cursor to the center of the gizmo and make sure its central portion is highlighted, refer to Figure 2-2. Next, drag the cursor up or down to scale the selected object. You can also perform non-uniform scaling using this tool. To do so, move the cursor over the axis along which you want to scale the object, refer to Figure 2-3, and then drag the cursor. Similarly, you can perform non-uniform scaling along the XY, YZ, or ZX plane by selecting the required plane and then dragging that plane, refer to Figure 2-4.

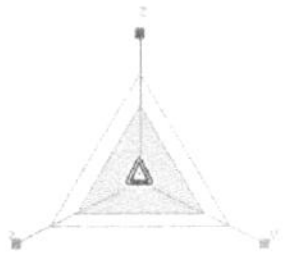

Figure 2-2 *The scale gizmo for uniform scaling*

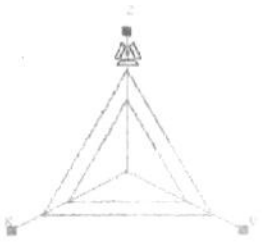

Figure 2-3 *Selecting the Z-axis for non-uniform scaling*

Figure 2-4 *Selecting the YZ plane for non-uniform scaling*

Select and Non-uniform Scale Tool

The **Select and Non-uniform Scale** tool is used to scale an object along a particular axis or plane non-uniformly. To scale the object, choose the **Select and Non-uniform Scale** tool from the Select and Scale flyout, and then select the object. Move the cursor over the X, Y, or Z axis along which you want to scale the object and drag the cursor to modify the shape of the object. Similarly, you can perform the non-uniform scaling along the XY, YZ, or ZX plane. You can also perform uniform scaling in the same manner, as described in the **Select and Uniform Scale** tool.

Select and Squash Tool

The **Select and Squash** tool is used to stretch and squash the object along the selected axis. To squash the object along one of the axes, choose the **Select and Squash** tool and select the object. Move the cursor over the axis along which you want to squash it. Next, drag the cursor. Similarly, you can squash the selected object along the XY, YZ, or ZX plane. To do so, move the cursor over the plane displayed between the two axes and drag the cursor to squash the object along that plane.

Note

*By default, the color of the X-axis, the Y-axis, and the Z-axis of the transform gizmos is red, green, and blue, respectively. When you move the cursor over any one of these axes, it gets activated and turns yellow. You can see the colors of these axes displayed at the bottom left corner of each viewport. Also, the colored axes are displayed while selecting an object using the **Select and Move**, **Select and Rotate**, or **Select and Scale** tool. Figures 2-5, 2-6, and 2-7 show the move gizmo, rotate gizmo, and scale gizmo, respectively.*

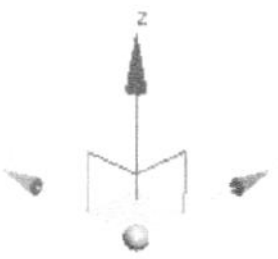

Figure 2-5 *The move gizmo*

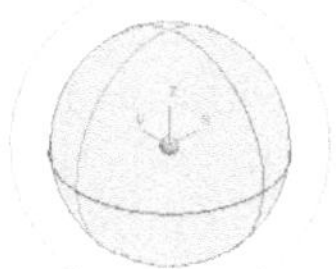

Figure 2-6 *The rotate gizmo*

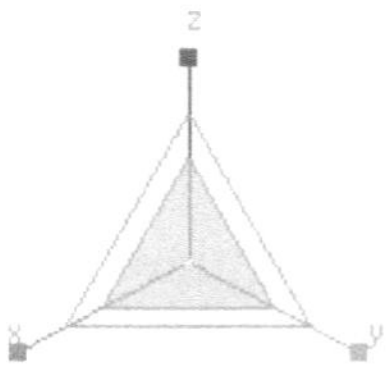

Figure 2-7 *The scale gizmo*

STANDARD PRIMITIVES

Autodesk 3ds Max has several basic three-dimensional geometric shapes which are known as standard primitives such as box, cone, sphere, cylinder, torus, tube, and so on. You can use these primitives to create simple 3D models such as a table, box, chair, and so on. All the standard primitives can be created dynamically using the mouse or by specifying the parameters in the **Keyboard Entry** rollout of the **Command Panel**.

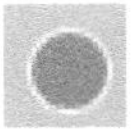

To create the standard primitives, choose **Create > Geometry** in the **Command Panel**. By default, the **Standard Primitives** option is selected in the drop-down list below the **Geometry** button. Now, activate the viewport in which you want to create the primitives.

Next, choose the corresponding tool from the **Object Type** rollout. In the following section, you will learn to create and modify the standard primitives using various tools available in the **Object Type** rollout.

Creating a Box

Menu bar:	Create > Standard Primitives > Box
Command Panel:	Create > Geometry > Standard Primitives> Object Type rollout > Box

To create a box, activate the viewport by clicking in it. Next, choose the **Box** tool from the **Object Type** rollout; the **Name and Color**, **Creation Method**, **Keyboard Entry**, and **Parameters** rollouts will be displayed, as shown in Figure 2-8. Press and hold the left mouse button in the viewport to specify the first corner of the box and then drag the cursor to define the length and width of the box. Release the left mouse button to get the desired length and width. Now, move the cursor up or down to define the height of the box. Click after you get the desired height; the box will be created, as shown in Figure 2-9.

Note

1. To view all the rollouts, position the cursor over any of the rollouts until you see the pan icon (hand). Next, press and hold the left mouse button and drag the cursor in the rollout.

2. This arrow sign on the left side of the rollout head indicates that the rollout is collapsed and the arrow sign indicates that the rollout is expanded.

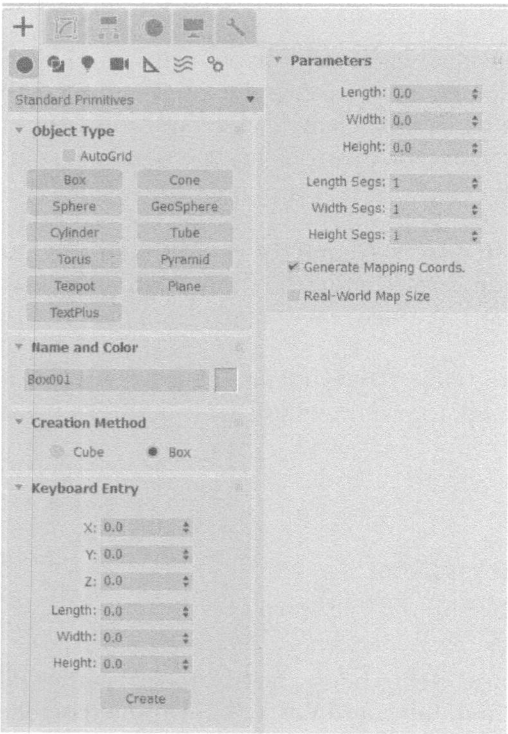

Figure 2-8 *Various rollouts to create a box*

Figure 2-9 *A box displayed in the viewport*

Creating a Cylinder

Menu bar:	Create > Standard Primitives > Cylinder
Command Panel:	Create > Geometry > Standard Primitives > Object Type rollout > Cylinder

The **Cylinder** tool is used to create a cylinder that can be sliced along its major axis. To create a cylinder, activate the required viewport by clicking in it and choose the **Cylinder** tool from the **Object Type** rollout; the **Name and Color**, **Creation Method**, **Keyboard Entry**, and **Parameters** rollouts will be displayed, as shown in Figure 2-10.

Press and hold the left mouse button to specify the center of the base of the cylinder and then drag the cursor to define the radius of the cylinder. Release the left mouse button. Next, move the cursor up or down to define the height of the cylinder. Click after you get the desired height; a cylinder will be created, as shown in Figure 2-11.

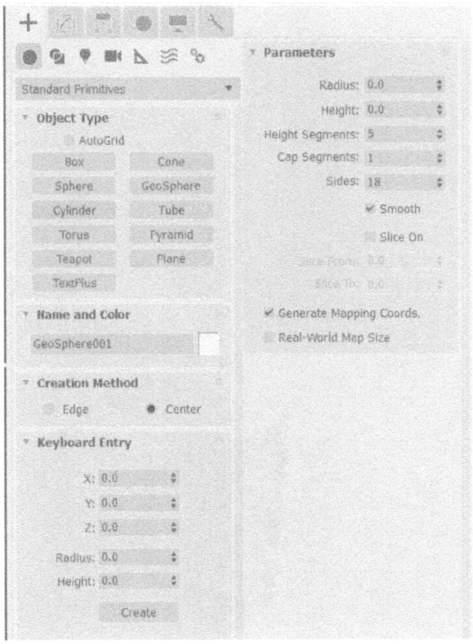

Figure 2-10 *Various rollouts to create a cylinder*

Figure 2-11 *A cylinder displayed in the viewport*

Creating a Tube

Menu bar:	Create > Standard Primitives > Tube
Command Panel:	Create > Geometry > Standard Primitives > Object Type rollout > Tube

The **Tube** tool is used to create round and prismatic tubes. The tube primitive is similar to the cylinder primitives with a hole in it. To create a tube, activate the required viewport and choose the **Tube** tool from the **Object Type** rollout; the **Name and Color**, **Creation Method**, **Keyboard Entry**, and **Parameters** rollouts will be displayed, as shown in Figure 2-12.

Press and hold the left mouse button to specify the center of the tube and then drag the cursor to define the first radius of the tube. It can be the inner or the outer radius of the tube. Next, release the left mouse button, move the cursor, and then click to define the second radius. Next, move the cursor up or down and click to get the desired height of the tube; a tube will be created, as shown in Figure 2-13.

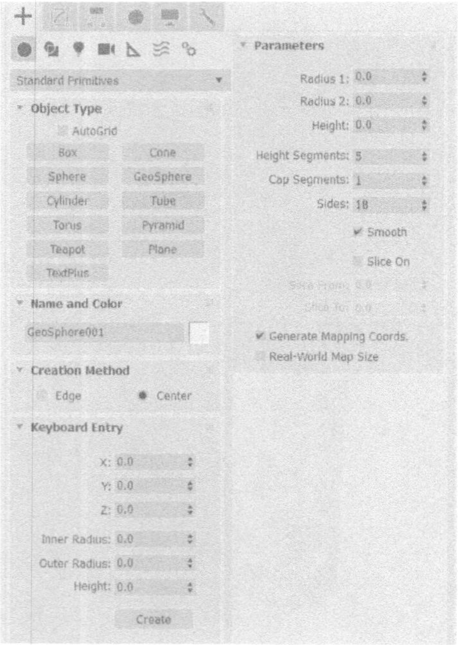

Figure 2-12 *Various rollouts to create a tube*

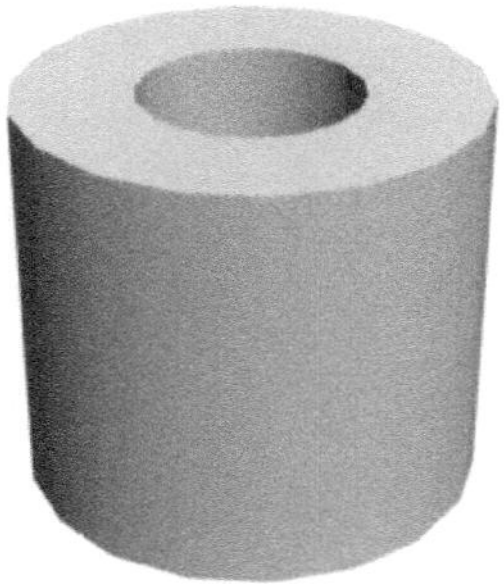

Figure 2-13 *A tube displayed in the viewport*

TUTORIAL

Tutorial 1 Creating a Coffee Table

In this tutorial, you will create 3D model of a coffee table and chairs, as shown in Figure 2-14, by using the standard primitives. **(Expected time: 30 min)**

Figure 2-14 *The model of coffee table and chairs*

The following steps are required to complete this tutorial:

a. Create the project folder.
b. Create top of the coffee table.
c. Create legs of the coffee table.
d. Create base of the coffee table
e. Create chairs.
f. Change background color of the scene.
g. Save and render the scene.

Creating the Project Folder

Before starting a new scene, it is recommended that you create a project folder. Creating a project folder helps you keep all the files of a project in an organized manner. Open the Windows Explorer and browse to the *Documents* folder. In this folder, create a new folder with the name *3dsmax2017*. The *3dsmax2017* folder will be the main folder and it will contain all the project folders that you will create while doing tutorials of this workbook. Now, you will create first project folder for Tutorial 1 of this chapter. To do so, you need to follow the steps given next:

1. Start Autodesk 3ds Max 2017.

2. If 3ds Max is already running, you need to reset it. To do so, choose **Reset** from the **Application** menu. If the current work is not saved, the **Autodesk 3ds Max 2017** message box is displayed. Next, choose the desired option from this message box; the **3ds Max** message box is displayed. Choose the **Yes** button from the message box; a new screen is displayed with default settings.

> **Note**
> *The **Reset** option is used to reset 3ds Max settings such as viewport configuration, snap settings, the **Material Editor** dialog box, background image, and so on. It also restores the startup defaults.*

3. Choose the **Application** button; the **Application** menu is displayed. Next, choose **Manage > Set Project Folder** from it; the **Browse For Folder** dialog box is displayed.

4. In the **Browse For Folder** dialog box, navigate to *\Documents\3dsmax2017*. Next, choose the **Make New Folder** button to create a new folder with the name *c02_tut1*. Next, choose the **OK** button to close the **Browse For Folder** dialog box.

5. Choose **Save** from the **Application** menu; the **Save File As** dialog box is displayed.

> **Note**
> *The scenes created in 3ds Max are saved with the .max extension. As the project folder is already created, the path \Documents\3dsmax2017\c02_tut1\scenes is displayed in the **Save in** drop-down list of the **Save File As** dialog box.*

> **Tip**
> *After setting the project when you open or save a scene, 3ds Max uses the **scenes** folder inside the project folder by default.*

6. Enter **c02tut1** in the **File name** edit box and then choose the **Save** button to close the dialog box.

Note

1. When you start 3ds Max, the last project that you have worked with is opened and an empty scene is created with the name Untitled.

2. You can also save a scene in old versions (2013 to 2017) of 3ds Max. To do so, select the desired option from the ***Save as type*** *drop-down in the* ***Save File As*** *dialog box.*

3. It is recommended that you frequently save the files while you are working on them by pressing the CTRL+S keys.

Autodesk 3ds Max creates a backup of the scene and saves changes periodically. In case of system failure, you can open the auto backup scene file and continue working on it. If you have not created the project folder on Windows 7, the default path to save the auto backup file is as follows: *C:\users\<username>\My Documents\3dsmax\autoback*. However, if you have created the project folder, the backup file will be saved in the *autoback* subfolder of the project folder. When 3ds Max is saving a file, the information, "**Autosave in progress... (Press ESC to cancel)**" is displayed in the prompt line at the bottom of the interface. If the size of the file is too large and it is taking time to save, you can press ESC to interrupt the saving process. You can also set the number of autoback files, their names, and backup time interval. To do so, choose **Customize > Preferences** from the menu bar; the **Preferences Settings** dialog box will be displayed. Next, choose the **Files** tab and then set the options as required in the **Auto Backup** area of dialog box.

Creating Top of the Coffee Table

In this section, you will create the top of the coffee table using the **Cylinder** tool and the **Tube** tool.

1. Choose **Create > Geometry** from the **Command Panel**; **Standard Primitives** is displayed in the drop-down list below the **Geometry** button. Also, the **Object Type** rollout is displayed in the **Command Panel**. Next, choose the **Tube** tool from the **Object Type** rollout.

2. Activate the Top viewport. Expand the **Keyboard Entry** rollout from the **Command Panel**. and enter the values in this rollout, as given next.

   Inner Radius : **60** Outer Radius : **80** Height: **4.2**

   Choose the **Create** button; a tube with the name *Tube001* is created in the viewports.

3. In the **Parameters** rollout, enter **48** in the **Sides** spinner; *Tube001* is displayed in the viewports, as shown in Figure 2-15.

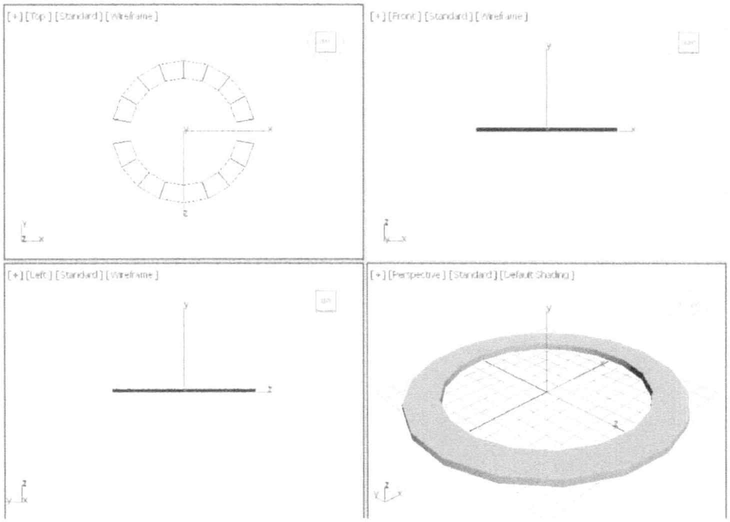

Figure 2-15 *The Tube001 displayed in viewports*

4. Choose the color swatch in the **Name and Color** rollout to change the color of *Tube001*; the **Object Color** dialog box is displayed. Now, using this dialog box, modify the color of *Tube001* to light brown.

5. Create another tube in the Top viewport by using the method described earlier. Enter the following values in the **Keyboard Entry** rollout:

   Z: **-0.5** Inner Radius : **60** Outer Radius : **40**
   Height: **4.2**

   Tube002 is displayed in the viewports, refer to Figure 2-16. In the **Parameters** rollout, enter **48** in the **Sides** spinner. Also, change the color of *Tube002* to light brown.

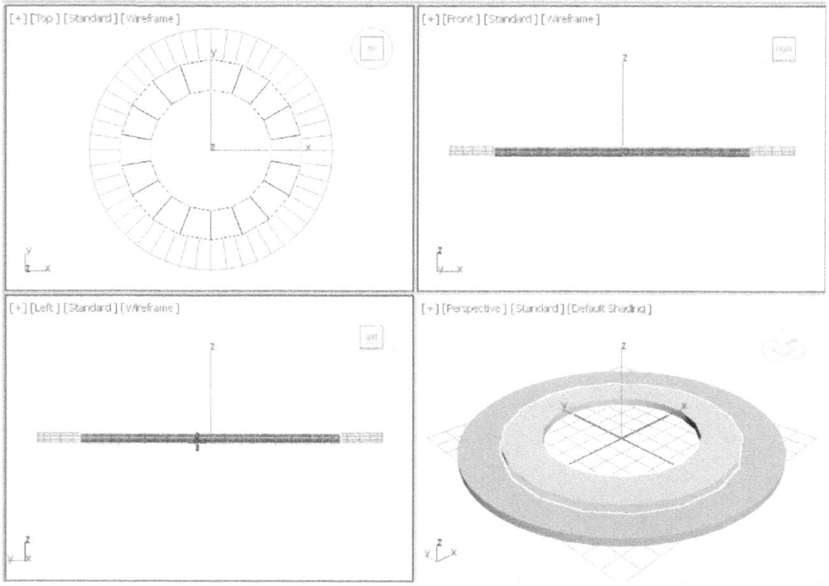

Figure 2-16 *The Tube002 displayed in viewports*

6. Make sure the Top viewport is activated and then choose the **Cylinder** tool from the **Object Type** rollout. Next, expand the **Keyboard Entry** rollout and enter the values as given next:

   Radius: **40** Height: **4.2**

   Choose the **Create** button; a cylinder is created in the viewports with the name *Cylinder001*, refer to Figure 2-17.

7. In the **Parameters** rollout, enter **48** in the **Sides** spinner. Also, change the color of *Cylinder001* to light brown.

8. Select *Tube001*, *Tube002*, and *Cylinder001* from the Scene Explorer and right-click; a quad menu is displayed. Choose **Add Selection To > New Group** from it; the **Group** dialog box is displayed. Enter **top** in this dialog box and choose **OK**; the *top* group is created with 3 objects in it.

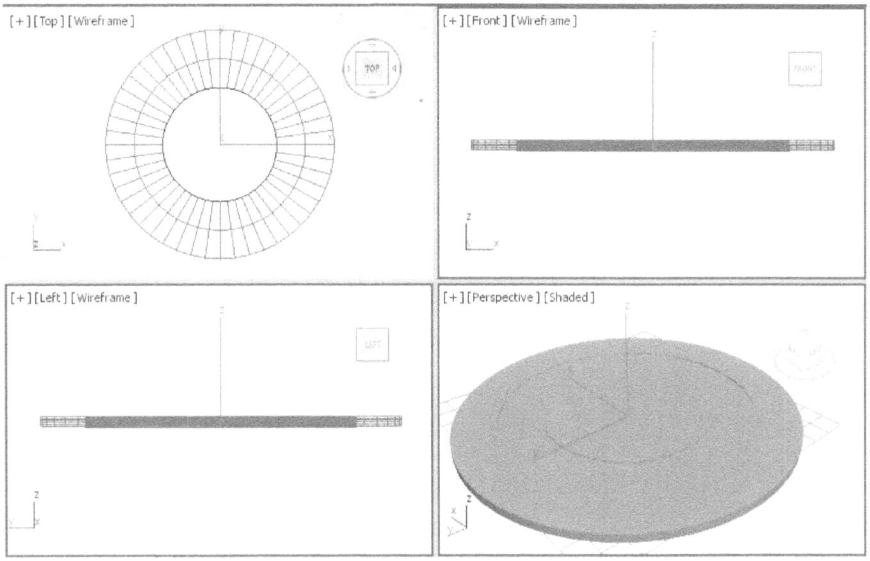

Figure 2-17 *The Cylinder001 displayed in the viewports*

Creating the Legs of the Coffee Table

In this section, you will create a leg of the table using the **Cylinder** tool and then clone it to create remaining legs..

1. Create cylinder in the Top viewport and enter the following values in the **Parameters** rollout:

   Radius: **5** Height: **150**

2. In the **Name and Color** rollout, enter **leg001** and press ENTER; the object is renamed. Next, change the color of *leg001* to light brown.

3. Align *leg001* with *Tube001* in viewports using the **Select and Move** tool, as shown in Figure 2-18.

   Next, you will create copies of *leg001*.

4. Activate the Top viewport and make sure *leg001* is selected. Next, move the cursor over the Y axis, press and hold the SHIFT key, and drag *leg001* upward. Now, release the left mouse button and the SHIFT key; the **Clone Options** dialog box is displayed, as shown in Figure 2-19.

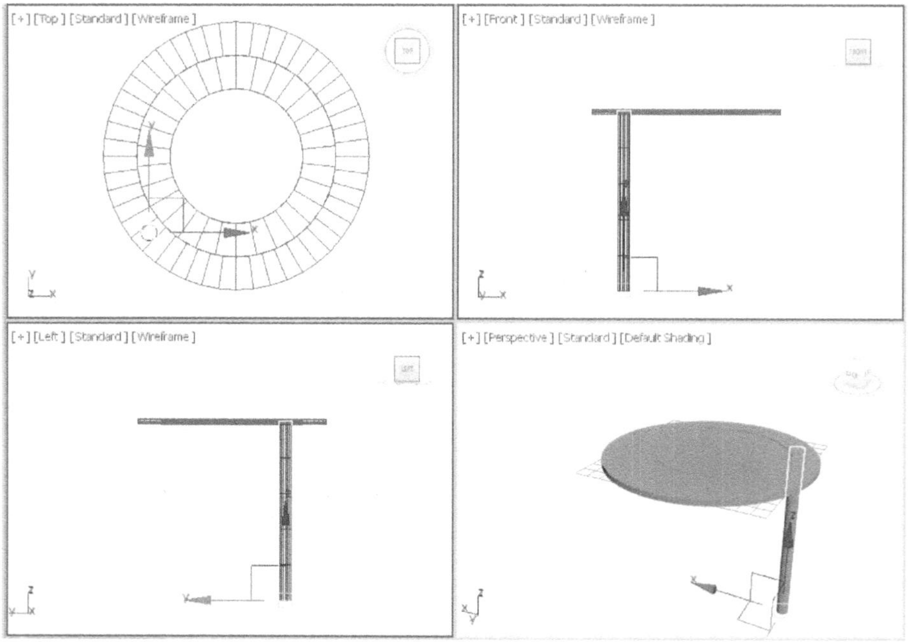

Figure 2-18 *The leg001 aligned in viewports*

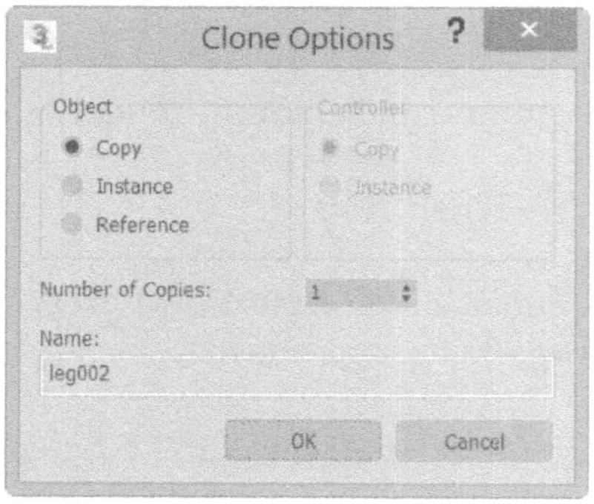

Figure 2-19 *The* ***Clone Options*** *dialog box*

5. Make sure the **Copy** radio button is selected in the **Object** area and 1 is displayed in the **Number of Copies** spinner of the **Clone Options** dialog box. Next, choose the **OK** button; *leg002* is created. Align *leg002* in the viewports, as shown in Figure 2-20.

Note

1. The cloned objects when created, if are not named, get automatically named in a sequence starting from the original object.

2. You can also clone an object by holding the SHIFT key while rotating or scaling it.

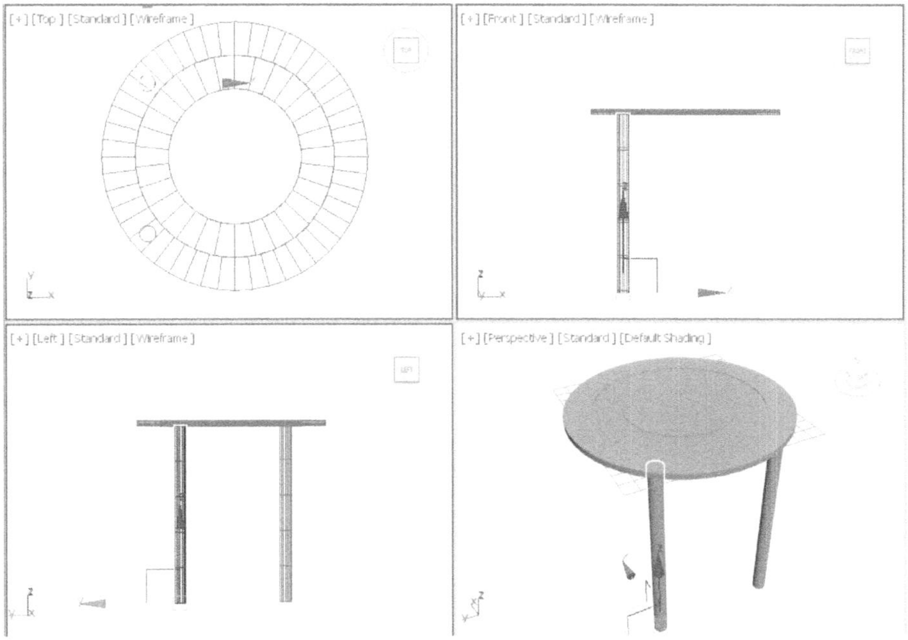

Figure 2-20 *The leg002 created after cloning leg001*

6. Select *leg001* and *leg002* from the Scene Explorer. In the Top viewport, move the cursor over the X axis, press and hold the SHIFT key, and drag *leg001* and *leg002* toward right until the value in the **X** spinner of the Coordinate display at the bottom of the screen becomes around **106**. Release the left mouse button and the SHIFT key; the **Clone Options** dialog box is displayed.

7. Make sure the **Copy** radio button is selected in the **Object** area and 1 is displayed in the **Number of Copies** spinner of the **Clone Options** dialog box. Next, choose the **OK** button; *leg003* and *leg004* are created, refer to Figure 2-21.

8. Select *leg001*, *leg002*, *leg003*, and *leg004* from the Scene Explorer and group them under the name *legs* as discussed earlier.

Creating Base of the Coffee Table

In this section, you will create base of the table by using the **Cylinder** tool.

1. Make sure the Top viewport is activated and then choose the **Cylinder** tool from the **Object Type** rollout. Next, expand the **Keyboard Entry** rollout and enter the values as given next:

   Z: **-125** Radius: **60** Height: **4.2**

   Choose the **Create** button; a cylinder is created in the viewports with the name *Cylinder002*.

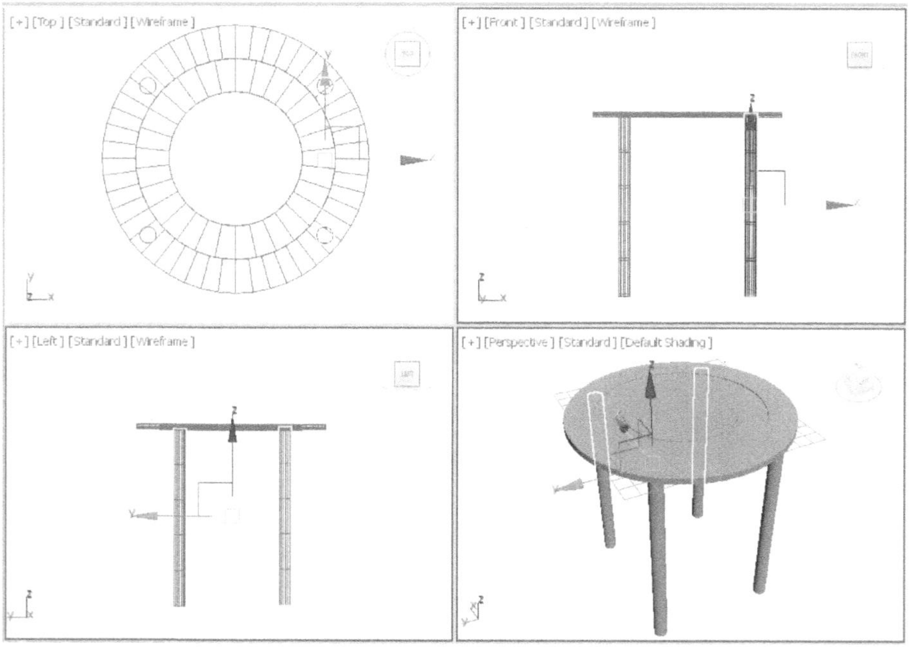

Figure 2-21 *The leg003 and leg004 created*

2. Rename it as *base01* and assign light brown color to it, refer to Figure 2-22.

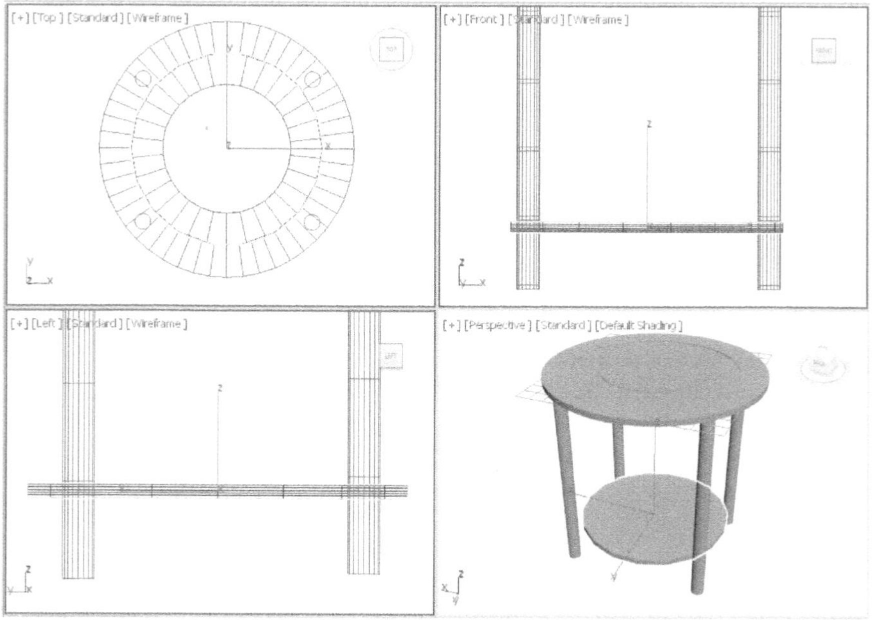

Figure 2-22 *The base01 created*

Next, you will create supports for *base01*.

3. Activate the Front viewport. Next, choose the **Cylinder** tool from the **Object Type** rollout.

4. Expand the **Keyboard Entry** rollout and enter the values as given next:

   Radius: **4** Height: **140**

   Choose the **Create** button; a cylinder is created in the viewports with the name *Cylinder002*.

5. In the **Name and Color** rollout, enter **support01** and press ENTER; the cylinder is named as *support01*. Also, change the color of *support01* to light brown.

6. Align *support01* using the **Select and Move** and **Select and Rotate** tools from the **Main Toolbar**, as shown in Figure 2-23.

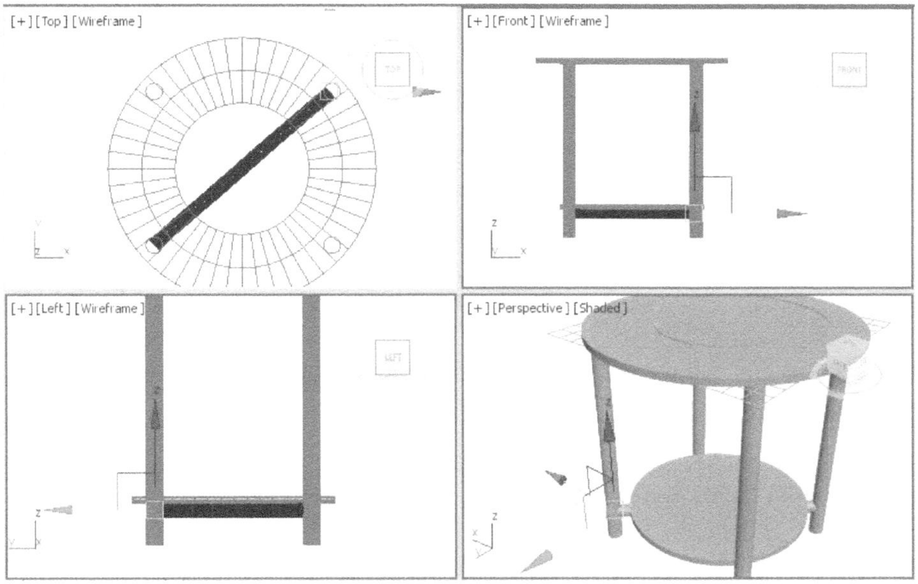

Figure 2-23 *Alignment of support01 in viewports*

7. In the Top viewport, copy *support01* by using the **Clone Options** dialog box as described earlier; *support002* is created. Now, align it with *base01* using the **Select and Move** and **Select and Rotate** tools, refer to Figure 2-24.

8. Select *base01*, *support01*, and *support002* from the Scene Explorer and group them under the name *base* as discussed earlier.

9. Select *base*, *top*, and *legs* groups from the Scene Explorer and group them under the name *table* as discussed earlier.

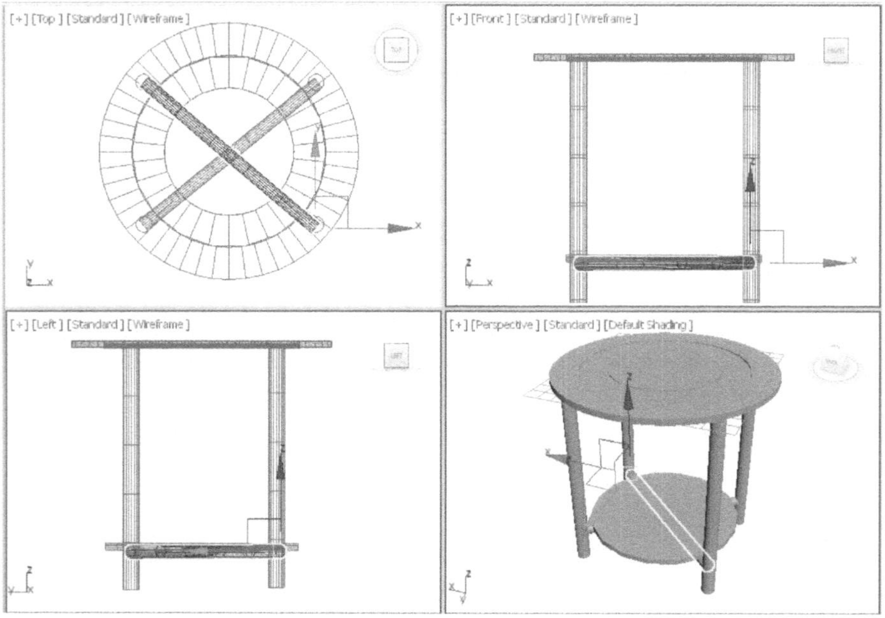

Figure 2-24 *Alignment of support002 with base01*

Creating Chair

In this section, you will create chair using the **Box** tool.

1. Select *table* from the Scene Explorer and right-click; a quad menu is displayed. Choose **Hide Selection** from it; *table* is hidden.

2. Activate the Top viewport. Next, choose the **Box** tool from the **Object Type** rollout and create a box. Now, choose the **Modify** tab from the **Command Panel**.

3. Rename the box as *base*. In the **Parameters** rollout, enter the values as given next:

   Length: **60** Width: **58** Height: **5**

   Next, you will create legs of the chair.

4. Repeat step 2 to create one more box. In the **Parameters** rollout, enter the values as given next:
   Length: **5** Width: **5** Height: **55**

5. Rename the box as *leg01*. Next, align *leg01* with *base,* as shown in Figure 2-25.

6. Create 3 copies of *leg01* as discussed earlier and align them, as shown in Figure 2-26.

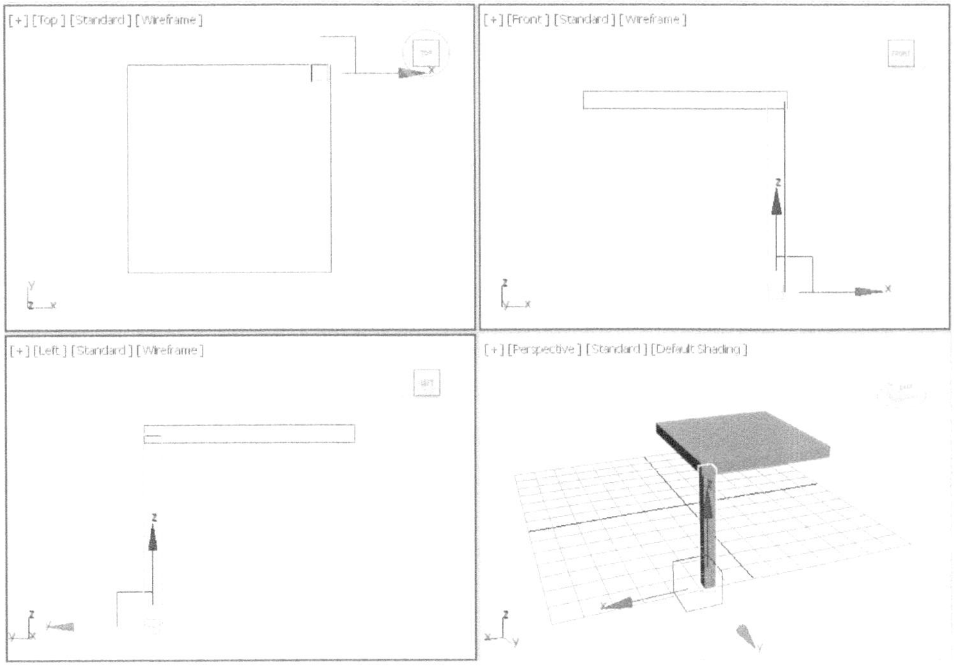

Figure 2-25 *Alignment of leg01 with base*

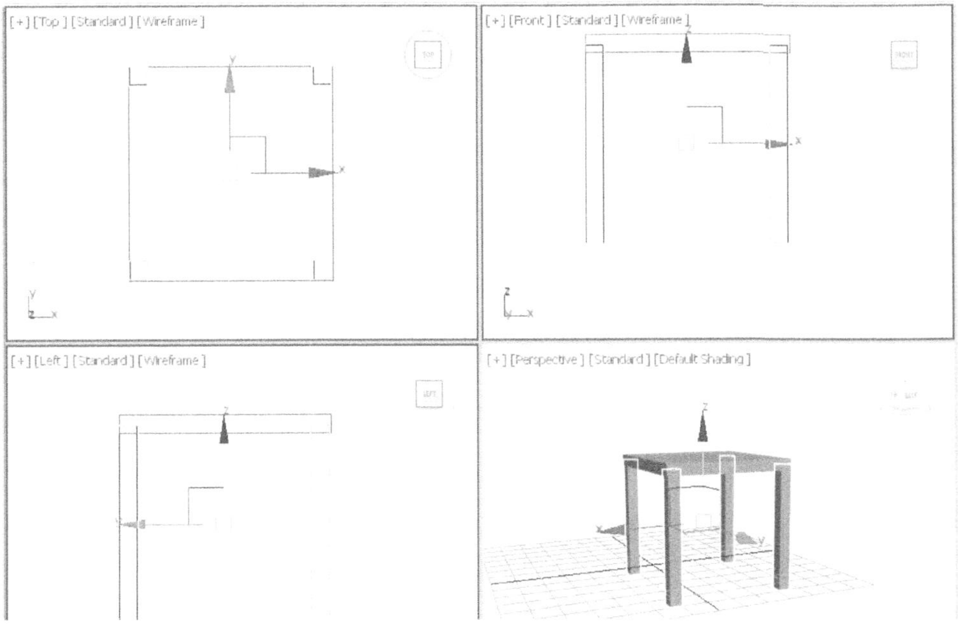

Figure 2-26 *Alignment of legs with base*

7. Select all the legs from the Scene Explorer and group them with the name *legs*.

   Next, you will create back of the chair.

8. Activate the Front viewport. Next, choose the **Box** tool from the **Object Type** rollout and create a box. Now, choose the **Modify** tab from the **Command Panel**.

9. In the **Parameters** rollout, enter the values as given next:

   Length : **9.5** Width : **65**

10. Rename the created object as *back1*. Next, align it with *base*, refer to Figure 2-27.

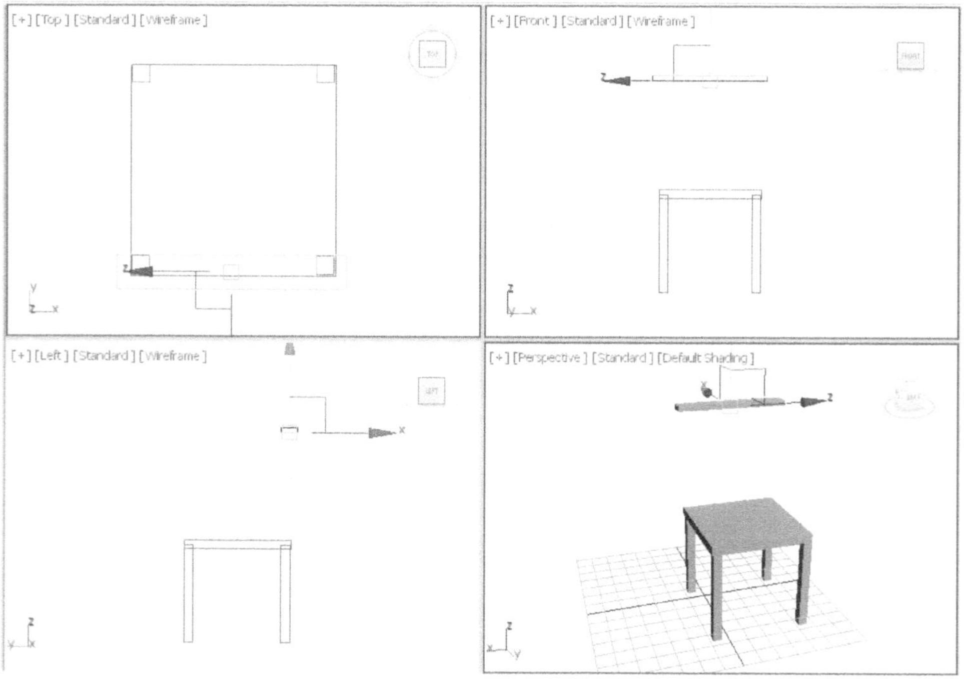

Figure 2-27 *Alignment of back1 with base*

   Note that you may need to adjust the position of *back1* at the time of alignment of objects in step 14.

11. Repeat step 8 to create one more box.

12. In the **Parameters** rollout, enter the values as given next:

   Length : **62** Width : **5.5** Height: **3.5**

13. Rename it as *back2*. Next, create 4 copies of *back2* and align them with *base* using the **Select and Move** and **Select and Rotate** tools, refer to Figure 2-28. Also, align *back1* with all these parts, as shown in Figure 2-29.

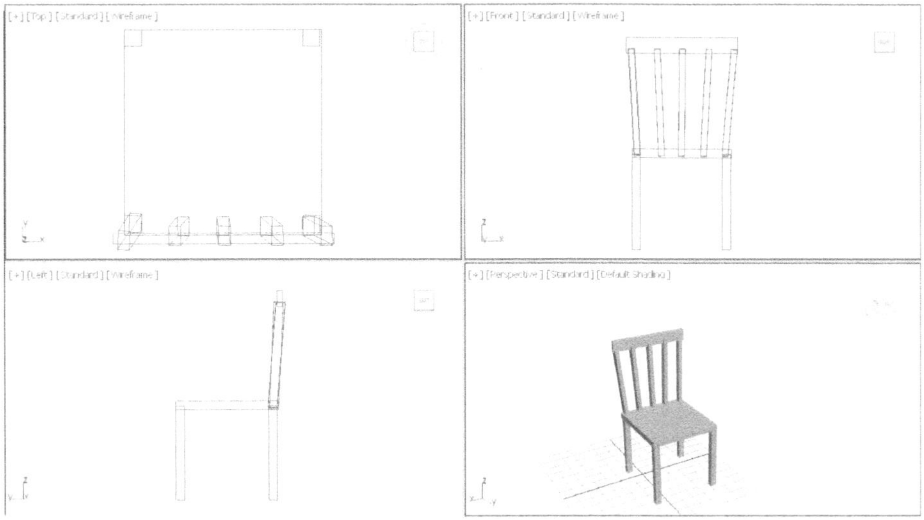

Figure 2-28 *Alignment of back2 and its copies*

14. Select *back1*, *back2*, *back3*, *back004*, *back005* and *back006* from the Scene Explorer and group them under the name *back* as discussed earlier.

    Next, you will create handrests for the chair.

15. Create a box in the Top viewport and enter the values in the **Parameters** rollout, as given next:

    Length: **45** Width: **5.2** Height: **2**

    Align it with *back2*, as shown in Figure 2-29.

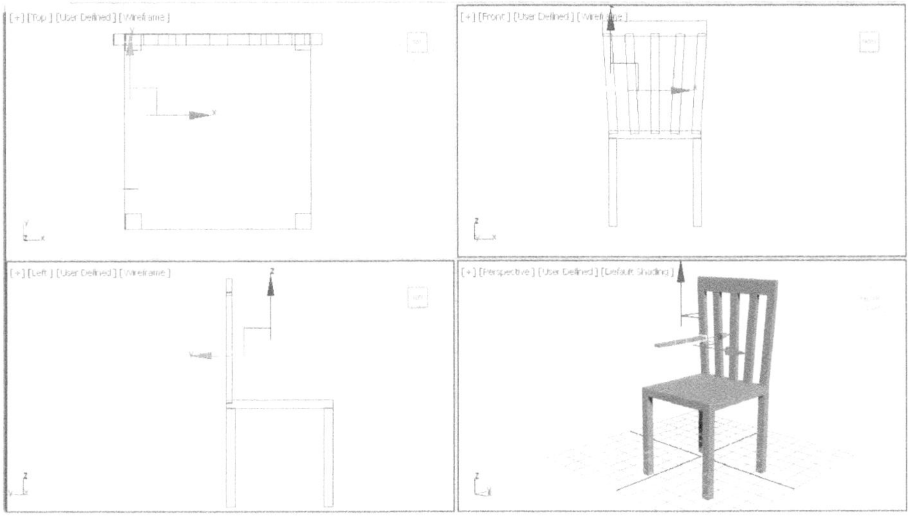

Figure 2-29 *Alignment of box with back2*

16. Create one more box in the Top viewport and enter the values in the **Parameters** rollout, as given next and then align it with the box created in step 16, as shown in Figure 2-30:

Length: **25** Width: **5.2** Height: **2**

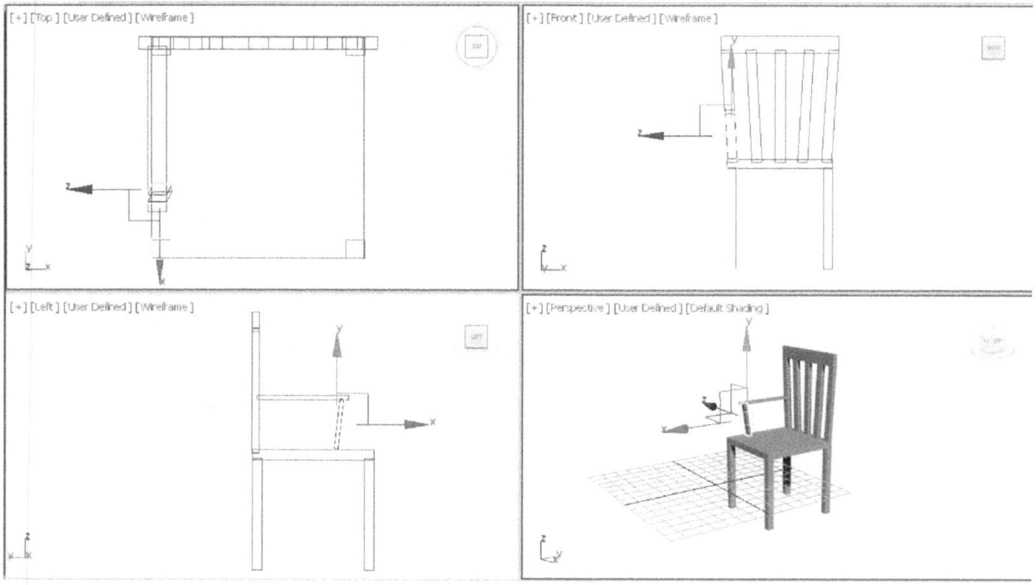

Figure 2-30 *Alignment of box in the viewports*

17. Create a copy of both the boxes created above and align them with *base,* as shown in Figure 2-31.

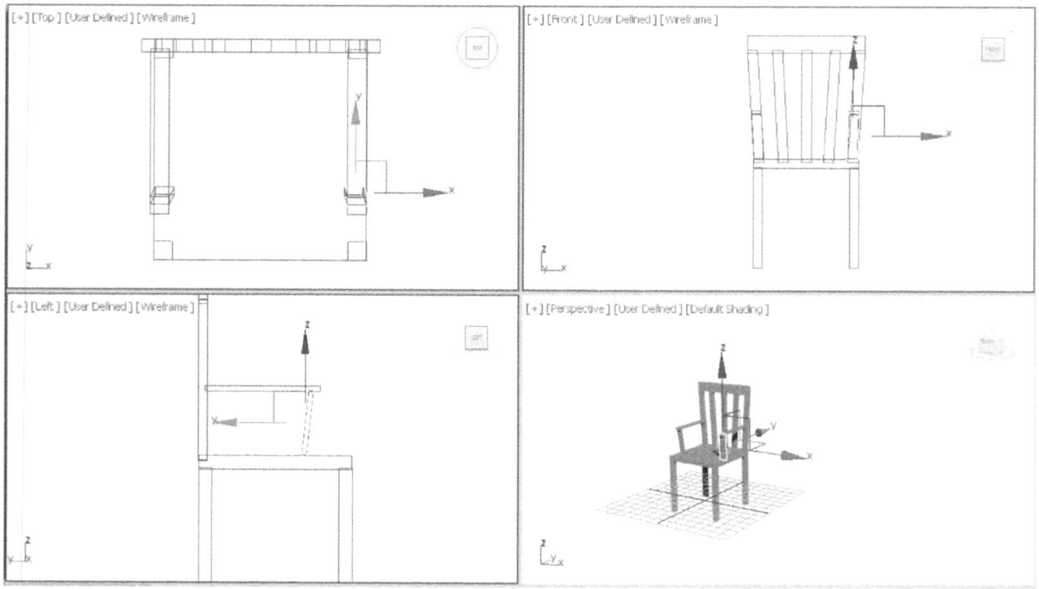

Figure 2-31 *Alignment of boxes in the viewports*

18. Create a group of four boxes created in above steps under the name *handrest.*

19. Select *back, base, legs,* and *handrest* from the Scene Explorer and group them with the name *chair.*

20. Right-click in the viewport and choose **Unhide All** from the quad menu displayed; *table* gets visible in the viewports.

21. Choose the **Select and Uniform Scale** tool from the **Main Toolbar** and scale *chair* in proportion to *table*.

22. Create a copy of *chair* and align *table* and chairs.

Changing Background Color of the Scene

In this section, you will change background color of the scene.

1. Choose **Rendering > Environment** from the menu bar; the **Environment and Effects** dialog box is displayed with the **Environment** tab chosen by default in this dialog box.

2. In the **Background** area of the **Common Parameters** rollout, choose the color swatch corresponding to the **Color** parameter; the **Color Selector: Background Color** dialog box is displayed. Select the white color and choose the **OK** button.

3. Close the **Environment and Effects** dialog box.

Saving and Rendering the Scene

In this section, you will save the scene that you have created and then render it. You can also view the final rendered image of this model by downloading the *c02_3dsmax_2017_rndr.zip* file from *www.cadsofttech.com* The path of the file is as follows: *Textbooks > Animation and Visual Effects > 3ds Max > Autodesk 3ds Max 2017 for Novices*

1. Choose **Save** from the **Application** menu.

2. Activate the Perspective viewport. Next, choose the **Render Production** tool from the **Main Toolbar**; the **Rendered Frame** window is displayed with the final output of the table, as shown in Figure 2-32.

Figure 2-32 *The final output after rendering*

EXERCISES

The rendered output of the models used in the following exercises can be accessed by downloading the *c02_3dsmax_2017_exr.zip* from *www.cadsofttech.com*. The path of the file is as follows: Textbooks > Animation and Visual Effects > 3ds Max > Autodesk 3ds Max 2017 for Novices.

Exercise 1

Create the model shown in Figure 2-33 using your own dimensions.

(Expected time: 15 min)

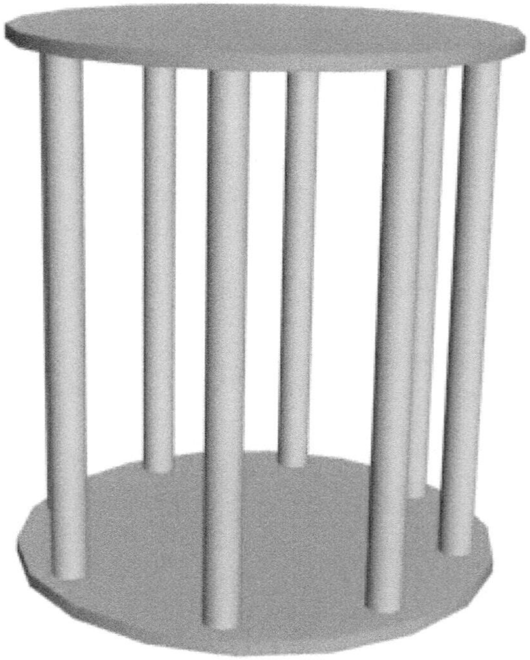

Figure 2-33 *The model to be created in Exercise 1*

Exercise 2

Create the nut model shown in Figures 2-34 and 2-35 using your own dimensions.

(Expected time: 15 min)

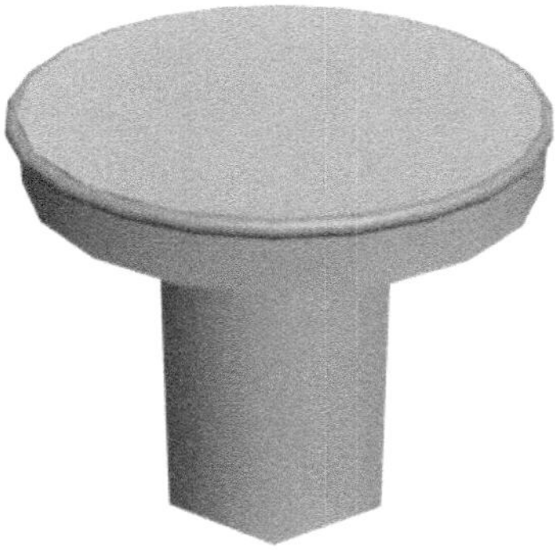

Figure 2-34 *The model of a nut (view 1)*

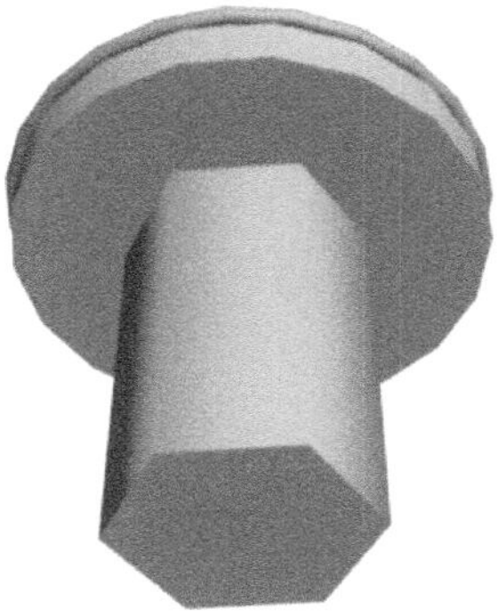

Figure 2-35 *The model of a nut (view 2)*

Exercise 3

Create the model of a table shown in Figure 2-36 using your own dimensions.

(Expected time: 15 min)

Figure 2-36 *The model of a table*

This page is intentionally left blank

Chapter 3

Extended Primitives

Learning Objectives

After completing this chapter, you will be able to:

- *Understand Bend modifier*
- *Understand Taper modifier*
- *Create and edit extended primitives*

INTRODUCTION

In this chapter, you will learn to create complex 3D shapes using extended primitives. Also, you will learn in detail the usage of the Bend and Taper modifiers and different snap controls.

MODIFIERS

In 3ds Max, modifiers are used to apply various types of functions or effects on an object. There are different types of modifiers that perform different functions about which you will learn in the later chapters. In this section, you will learn about the **Bend** and **Taper** modifiers.

Bend Modifier

The **Bend** modifier is used to bend an object upto 360 degrees along the X, Y, or Z axis. To apply the **Bend** modifier on an object, select the object and choose **Modifiers > Parametric Deformers > Bend** from the menu bar. Alternatively, you can select the object in the viewport and choose the **Modify** tab from the **Command Panel**; the **Modifiers List** drop-down list will be displayed. Select **OBJECT-SPACE MODIFIERS > Bend** from the **Modifier List** drop-down list; the **Bend** modifier will be displayed in the modifier stack, refer to Figure 3-1. Also, the **Parameters** rollout will be displayed in the Modify panel, as shown in Figure 3-2.

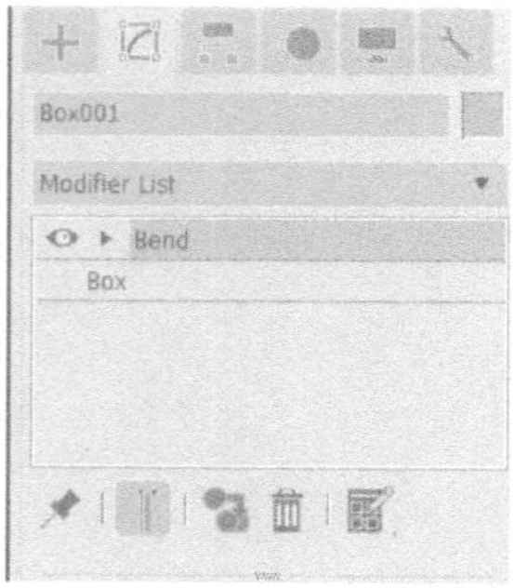

*Figure 3-1 The **Bend** modifier displayed in the modifier stack*

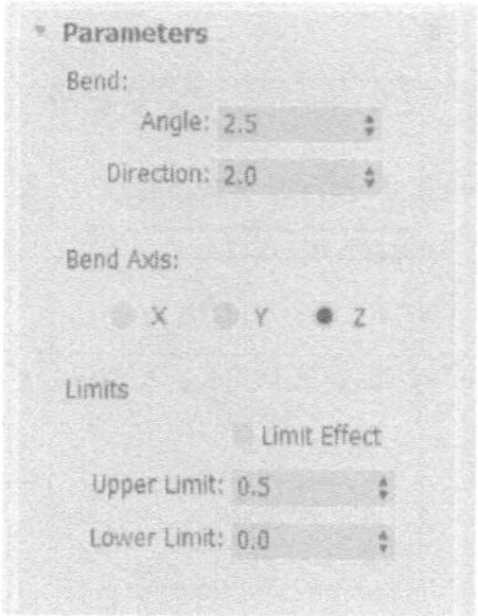

*Figure 3-2 The **Parameters** rollout of the **Bend** modifier*

Taper Modifier

The **Taper** modifier is used to taper an object by scaling it on both ends. After applying this modifier, one side of the object will be scaled up and another will be scaled down. To apply the **Taper** modifier, choose **Modifiers > Parametric Deformers > Taper** from the menu bar; the **Taper** modifier will be displayed in the modifier stack, refer to Figure 3-3. Also, the **Parameters** rollout will be displayed in the Modify panel, as shown in Figure 3-4.

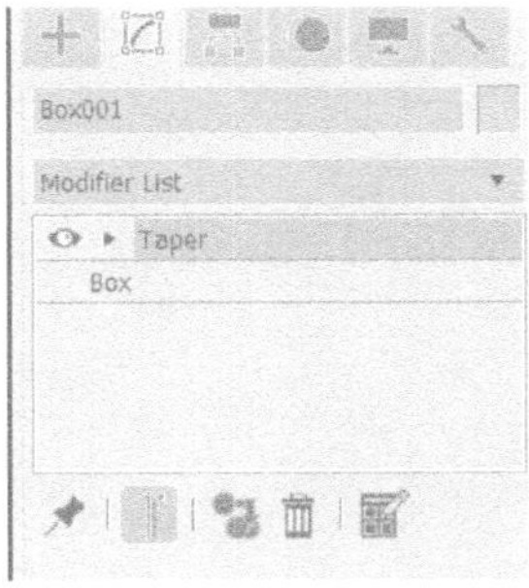

*Figure 3-3 The **Taper** modifier in the modifier stack*

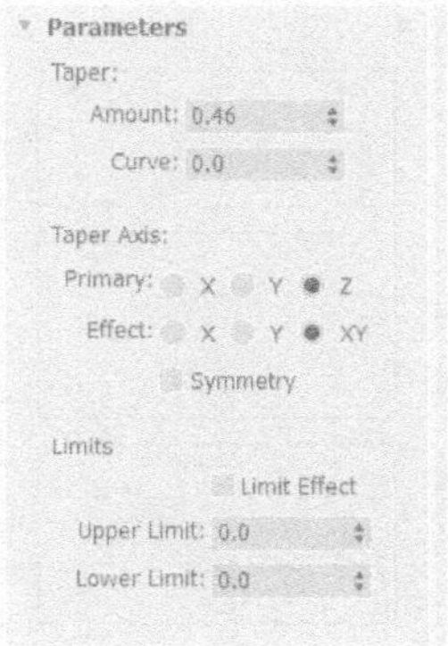

*Figure 3-4 The **Parameters** rollout of the **Taper** modifier*

EXTENDED PRIMITIVES

In 3ds Max, there are some complex 3D geometric shapes, which can be used to create 3D objects. These shapes are known as extended primitives. The extended primitives are chamfer box, spindle, hedra, and so on.

To create an extended primitive, you need to choose **Create > Geometry** from the **Command Panel**; a drop-down list will be displayed with the **Standard Primitives** option selected by default. Now, select the **Extended Primitives** option from the drop-down list. Activate the viewport in which you want to create the primitives and then choose the corresponding tool from the **Object Type** rollout. In this section, you will learn to create and modify the extended primitives using various tools available in the **Object Type** rollout.

Creating a Chamfer Box

Menu bar:	Create > Extended Primitives > Chamfer Box
Command Panel:	Create > Geometry > Extended Primitives > Object Type rollout > ChamferBox

A chamfer box is similar to a box in the standard primitives with the only difference is that it has beveled or rounded edges. To create a chamfer box, activate the viewport and choose the **ChamferBox** tool from the **Object Type** rollout; the **Name and Color**, **Creation Method**, **Keyboard Entry**, and **Parameters** rollouts will be displayed, as shown in Figure 3-5.

Now, press and hold the left mouse button at a point in the viewport for defining the first corner of the chamfer box, and drag the cursor to define its length and width. Release the left mouse button at a point in the viewport for defining the first corner of the chamfer box, and drag the cursor to define its length and width. Release the left mouse button to get the desired length and width. Next, move the cursor up or down to define its height, and click on the viewport to get the desired height. Move the cursor to define the fillet and click on the viewport to get the fillet or chamfer; a chamfer box will be created.

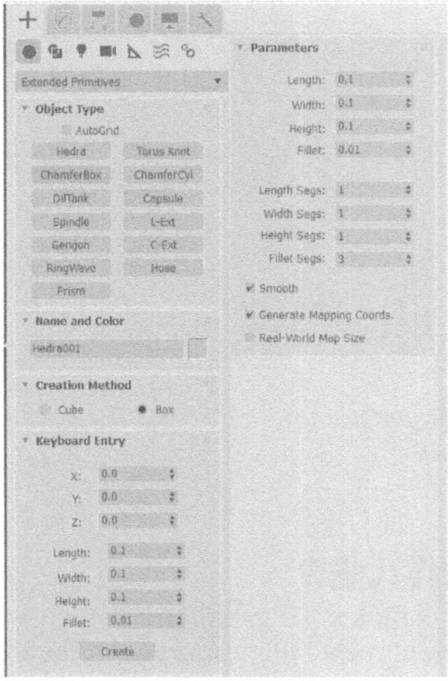

Figure 3-5 *Various rollouts to create a chamfer box*

Creating a Hose

Menu bar:	Create > Extended Primitives > Hose
Command Panel:	Create > Geometry > Extended Primitives > Object Type rollout > Hose

A hose is an extended primitive that can link two objects. I t reacts to the movement of linked objects. To create a hose, activate the viewport and choose the **Hose** tool from the **Object Type** rollout; the **Name and Color** and **Hose Parameters** rollouts will be displayed. Now, press and hold the left mouse button in the viewport to specify the center of the hose and drag the cursor to define its radius. Release the left mouse button and move the cursor up or down to define the height of the hose. Click on the screen to get the desired height of the hose; a hose will be created, as shown in Figure 3-6.

Figure 3-6 *A hose displayed in the viewport*

Tutorial 1

In this tutorial, you will create the model of a chair, as shown in Figure 3-7, using the extended primitives and modifiers. **(Expected time: 40 min)**

Figure 3-7 *The model of a chair*

The following steps are required to complete this tutorial:

a. Create the project folder.
b. Create seat support.
c. Create back support.
d. Create spring support.
e. Create leg support.
f. Create legs.
g. Create the rollers for legs.
h. Create right hand support.
i. Create left hand support.
j. Save and render the scene.

Creating the Project Folder

Create a new project folder with the name *c03_tut2* at *\Documents\3dsmax2017* and then save the file with the name **c03tut2**, as discussed in Tutorial 1 of Chapter 2.

Creating Seat Support

In this section, you need to create a chamfer box primitive to create seat support of the chair.

1. Activate the Top viewport. Choose **Create > Geometry** from the **Command Panel**; the **Standard Primitives** option is displayed by default in the drop-down list. Select the **Extended Primitives** option from the list and then choose the **ChamferBox** tool from the **Object Type** rollout.

2. Expand the **Keyboard Entry** rollout and set the values as follows:

   Length: **90.0** Width: **100.0** Height: **17.0**
   Fillet: **10.0**

3. Choose the **Create** button; a chamfer box is created.

4. Make sure the chamfer box is selected. In the **Parameters** rollout, set the value **10** in the **Fillet Segs** spinner; the chamfer box is modified, as shown in Figure 3-8.

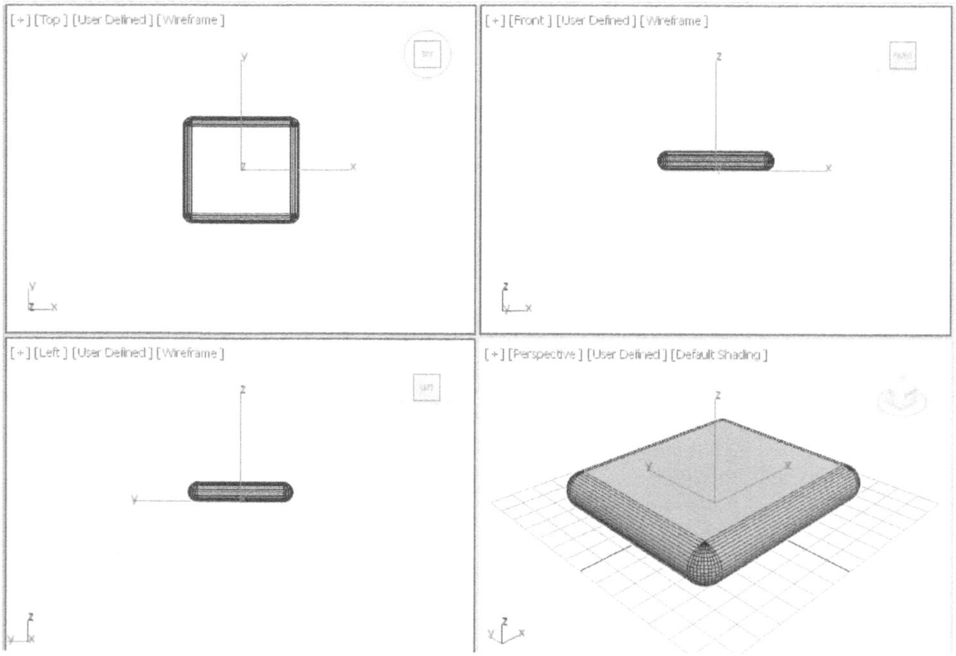

Figure 3-8 *The chamfer box created for seat support of the chair*

5. In the **Name and Color** rollout, enter **seat support** as the name of the chamfer box and press ENTER.

6. Use the color swatch to modify the color of *seat support* by entering the values as follows:

   Red: **88** Green: **199** Blue: **225**

Creating Back Support

In this section, you will use the **ChamferBox** tool from the extended primitives to create back support for the chair.

1. Activate the Front viewport and make sure that the **ChamferBox** tool is chosen.

2. Expand the **Keyboard Entry** rollout and set the values as follows:

   X: **0.0** Y: **65.0** Z: **-65.0** Length: **90.0**
   Width: **100.0** Height: **19.633** Fillet: **10.0**

3. Choose the **Create** button; a chamfer box is created. In the **Parameters** rollout, make sure the value in the **Fillet Segs** spinner is **10** and enter **back support** as the name of the chamfer box in the **Name and Color** rollout, refer to Figure 3-9.

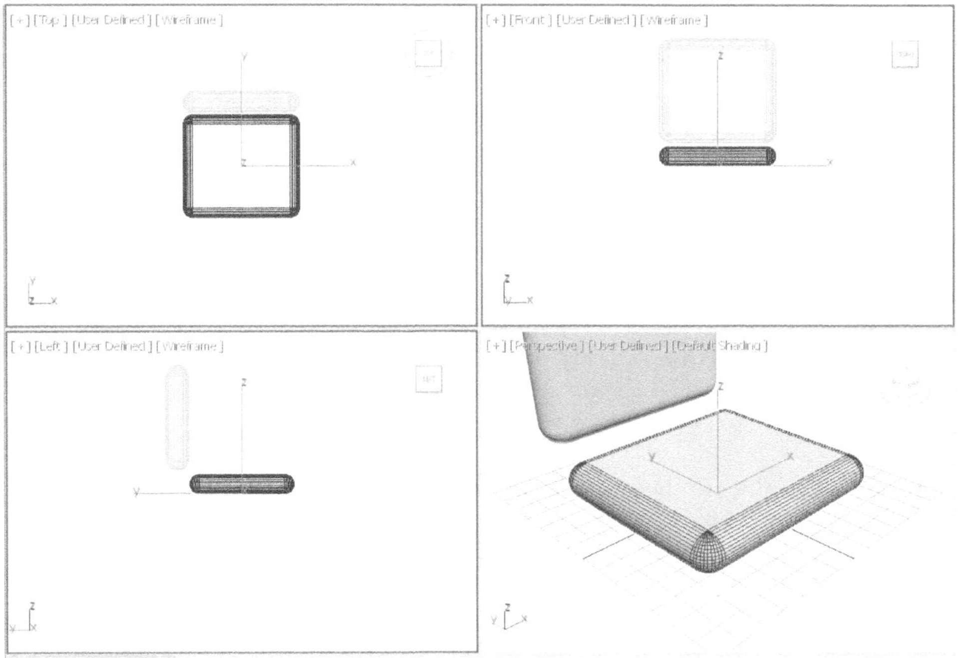

Figure 3-9 The back support geometry in viewports

4. Choose the color swatch and assign the same color to *back support* that you assigned to *seat support*.

5. Choose the **Zoom Extents All** tool to display the objects properly in all viewports.

   Next, you need to create the spring support to join *seat* and *back support*.

Creating Spring Support

In this section, you will create spring support by using the **Hose** tool from the extended primitives.

1. Activate the Top viewport. Choose **Create > Geometry** from the **Command Panel**. Make sure that the **Extended Primitives** option is selected in the drop-down list and then choose the **Hose** tool from the **Object Type** rollout.

2. In the Top viewport, press and hold the left mouse button and drag the cursor to specify the radius of the hose, and then release the left mouse button. Next, move the cursor up to specify the height of the hose and click on the viewport; a hose is created.

3. In the **Name and Color** rollout, enter **spring support** as the name of the hose and press ENTER. Also, modify the color of *spring support* to black.

4. In the **Hose Parameters** rollout, set the values of *spring support* as follows:

   End Point Method Area
   Make sure that the **Free Hose** radio button is selected.

Free Hose Parameters Area
Height: **25.815**

Common Hose Parameters Area
Cycles: **9**
Use the default values for the other options in this area.

Smoothing Area
Make sure the **All** radio button is selected.

Hose Shape Area
Make sure the **Round Hose** radio button is selected.

Diameter: **15.0**   Sides: **8**

After setting the values in the **Hose Parameters** rollout, align *spring support* in the viewports, as shown in Figure 3-10.

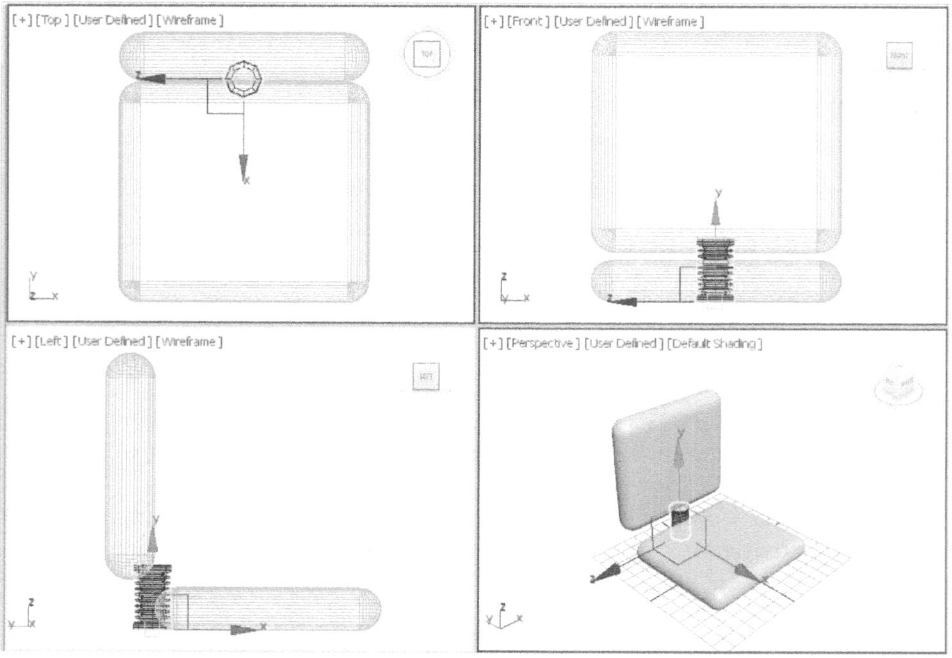

Figure 3-10 *The spring support geometry in viewports*

Next, you will apply the **Bend** modifier to *spring support* to join *seat* and *back support*.

5. Make sure *spring support* is selected and then choose **Modifiers > Parametric Deformers > Bend** from the menu bar; the **Bend** modifier is displayed in the modifier stack. Also, the **Parameters** rollout is displayed in the **Modify** tab.

6. In the **Parameters** rollout of the **Bend** modifier, set the values as follows:

   Bend Area
   Angle: **90.0** Direction: **90**

   Bend Axis area
   Make sure the **Z** radio button is selected.

7. After entering the values, *spring support* is displayed,as shown in Figure 3-11.

   Next, you will align *spring support*.

8. Activate the Left viewport and make sure that *spring support* is selected. Choose the **Zoom Extents Selected** tool; *spring support* is zoomed in the Left viewport, as shown in Figure 3-12.

9. Right-click on the **Select and Rotate** tool; the **Rotate Transform Type-In** dialog box is displayed. In the **Offset: Screen** area, set **90** in the **Z** spinner and press ENTER; *spring support* is rotated, as shown in Figure 3-13. Now, close the **Rotate Transform Type-In** dialog box.

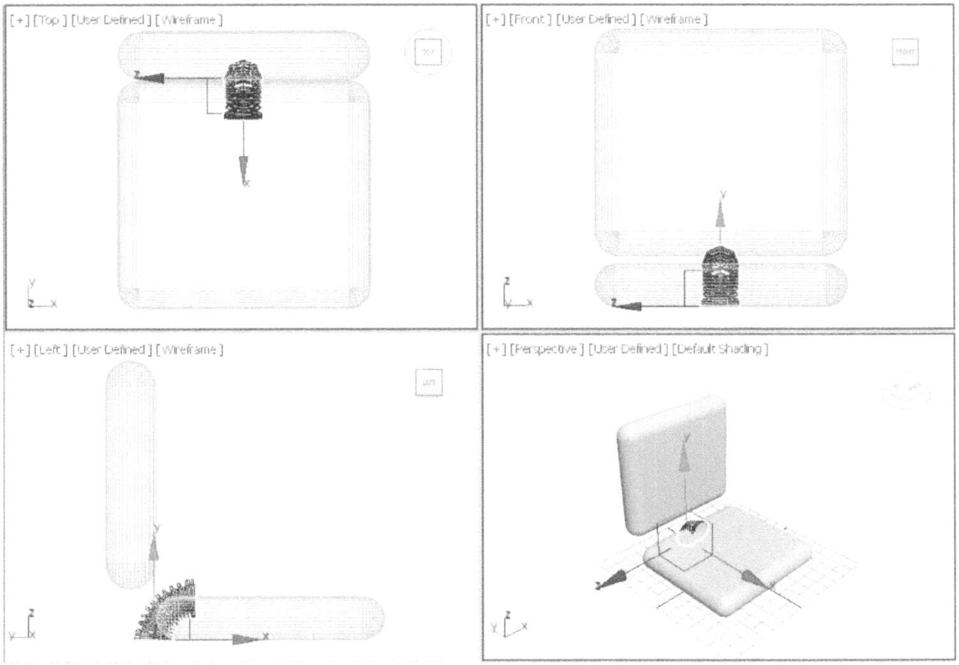

Figure 3-11 *The spring support geometry after applying the **Bend** modifier*

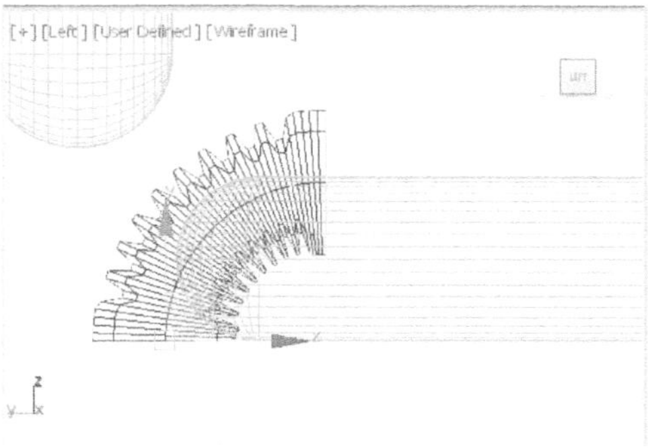

Figure 3-12 *The spring support geometry zoomed in the Left viewport*

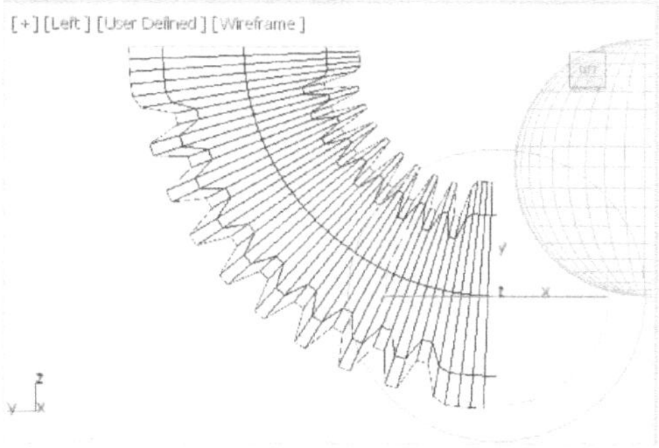

Figure 3-13 *The spring support geometry rotated in the Left viewport*

10. Choose the **Select and Move** tool and align *spring support*, as shown in Figure 3-14.

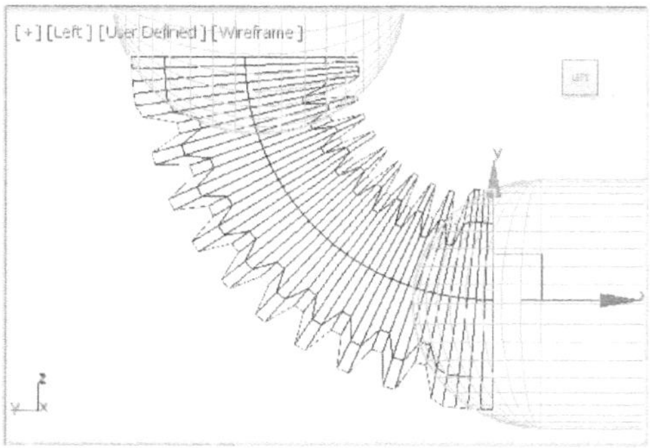

Figure 3-14 *Alignment of spring support in the Left viewport*

11. Choose the **Zoom Extents All** tool to display all the objects properly in the viewports, refer to Figure 3-15.

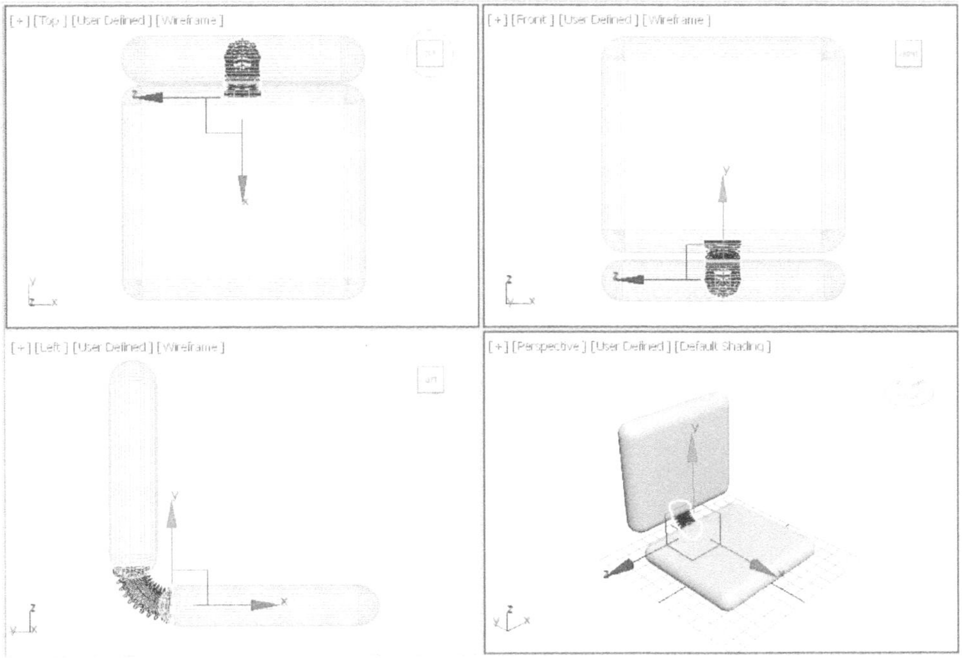

Figure 3-15 Alignment of spring support in the viewports

Creating Leg Support

Now, you need to create leg support. You will use the **Hose** tool from the extended primitives to create it.

1. Activate the Top viewport and then create a hose object as discussed earlier.

2. In the **Name and Color** rollout, enter **leg support** as the name of the hose and press ENTER. Also, modify the color of *leg support* to black.

3. In the **Hose Parameters** rollout, set the values of *leg support* as follows:

 End Point Method Area
 Make sure that the **Free Hose** radio button is selected.

 Free Hose Parameters Area
 Height: **42.0**

 Common Hose Parameters Area
 Make sure that the **Flex Section Enable** check box is selected.

 Starts: **22.0** Ends: **82.0**
 Cycles: **10**

Make sure the value in the Diameter spinner is **-20.0**.

Smoothing Area
Make sure the **All** radio button is selected.
Hose Shape Area
Make sure the **Round Hose** radio button is selected.

Diameter: **17.0** Sides: **8**

4. Choose the **Select and Move** tool and align *leg support* in the viewports, as shown in Figure 3-16.

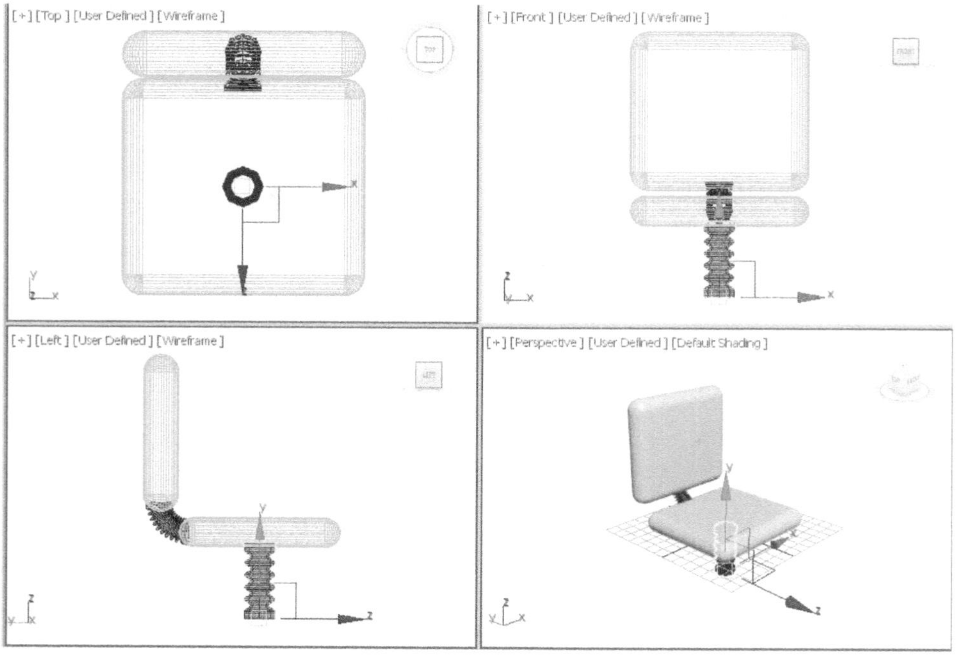

Figure 3-16 *Alignment of leg support in the viewports*

Creating Legs

In this section, you will use the **ChamferBox** tool from **Extended Primitives** to create legs of the chair.

1. Activate the Top viewport and choose the **ChamferBox** tool.

2. Expand the **Keyboard Entry** rollout and set the values as follows:

X: **-29** Y: **0.0** Z: **-42.0** Length: **5.748**
Width: **44.0** Height: **7.0** Fillet: **2.0**

3. Choose the **Create** button; a chamfer box is created.

4. Make sure the chamfer box is selected. In the **Parameters** rollout, set the values as follows:

Length Segs: **10** Width Segs: **10** Height Segs: **10**
Make sure the value in the Fillet Segs is **10**.
After entering the values, the chamfer box is modified.

5. In the **Name and Color** rollout, enter **leg01** as the name of the chamfer box and press ENTER. Also, modify its color to black.

6. Choose the **Select and Move** tool and align *leg01* with *leg support* in the viewports, as shown in Figure 3-17.

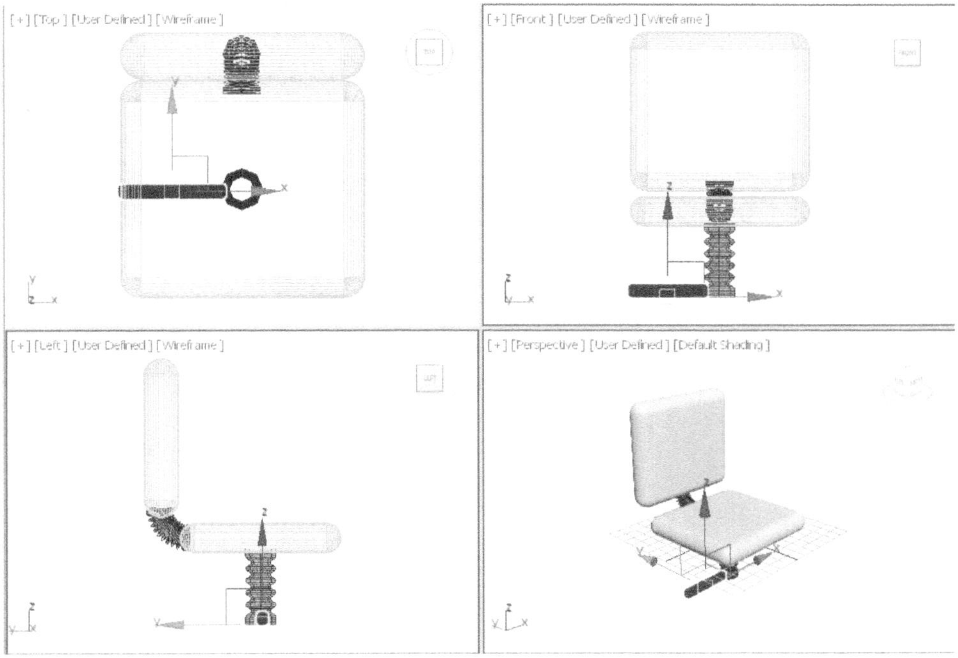

Figure 3-17 *Alignment of leg01 in the viewports*

Now, you need to apply the **Taper** modifier to *leg01* to taper it.

7. Make sure that *leg01* is selected, and then choose the **Zoom Extents All Selected** tool; *leg01* is zoomed in all viewports, as shown in Figure 3-18.

8. Choose **Modifiers > Parametric Deformers > Taper** from the menu bar; the **Taper** modifier is displayed in the modifier stack. Also, the **Parameters** rollout is displayed in the Modify tab.

9. In the **Parameters** rollout of the **Taper** modifier, set the values as given below:

Taper Area
Amount: **0.53**

Taper Axis Area
Select the **X** radio button in the **Primary** group and make sure the **ZY** radio button is selected in the **Effect** group.

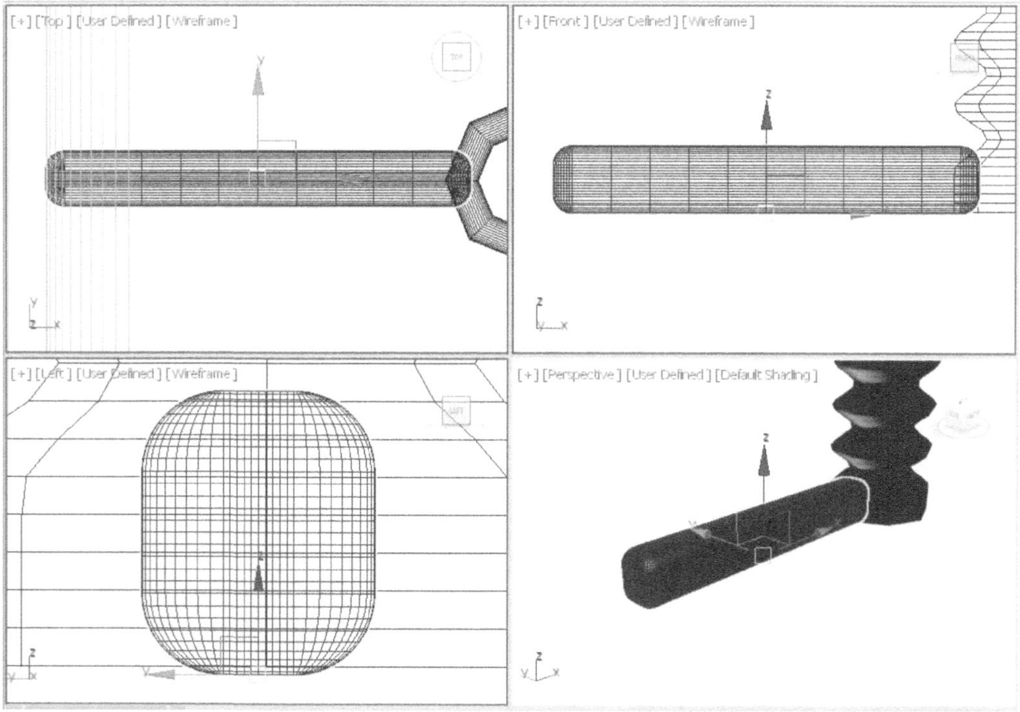

Figure 3-18 *The leg01 geometry after invoking the* ***Zoom Extents All Selected*** *tool*

10. After entering the values, *leg01* is tapered, as shown in Figure 3-19.

Next, you need to create rollers.

Creating the Rollers for Legs

In this section, you will create rollers. You will use the chamfer box and chamfer cylinder primitives to create them.

1. Activate the Top viewport and choose the **ChamferBox** tool.

2. Expand the **Keyboard Entry** rollout and set the values as follows:

   X: **-40** Y: **0.0** Z: **-20.0** Length: **7.032**
   Width: **14.816** Height: **1.799** Fillet: **1.0**

3. Choose the **Create** button; a chamfer box is created.

4. Make sure that the chamfer box is selected. In the **Parameters** rollout, make sure the value in the Length Segs, Width Segs, Height Segs, and Fillet Segs spinners is **10**.

5. In the **Name and Color** rollout, enter **box01** as the name of the chamfer cylinder. Also, modify its color to black.

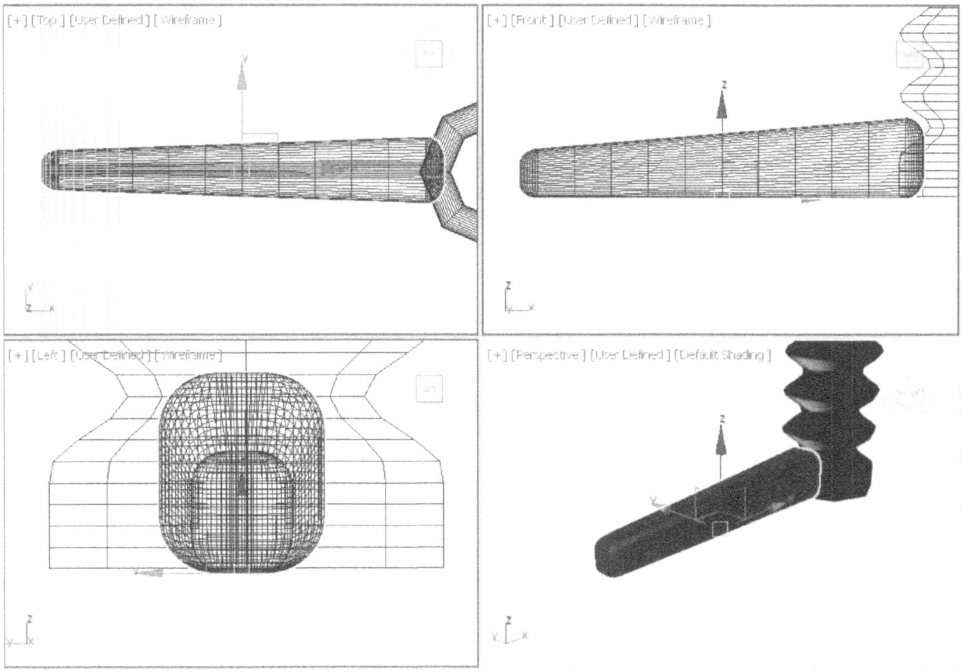

*Figure 3-19 The leg01 geometry displayed in viewports after applying the **Taper** modifier*

Note

*To view box01 in viewports, you need to adjust the view by invoking the **Zoom** and **Pan View** tools.*

Next, you will apply the **Bend** modifier to *box01*.

6. Choose **Modifiers > Parametric Deformers > Bend** from the menu bar; the **Bend** modifier is displayed in the modifier stack. Also, the **Parameters** rollout is displayed in the Modify panel.

7. In the **Parameters** rollout of the **Bend** modifier, set the values as follows:

   Bend Area
   Angle: **173.5**
   Bend Axis Area
   Select the **X** radio button.

8. After applying the **Bend** modifier, *box01* is displayed, as shown in Figure 3-20.

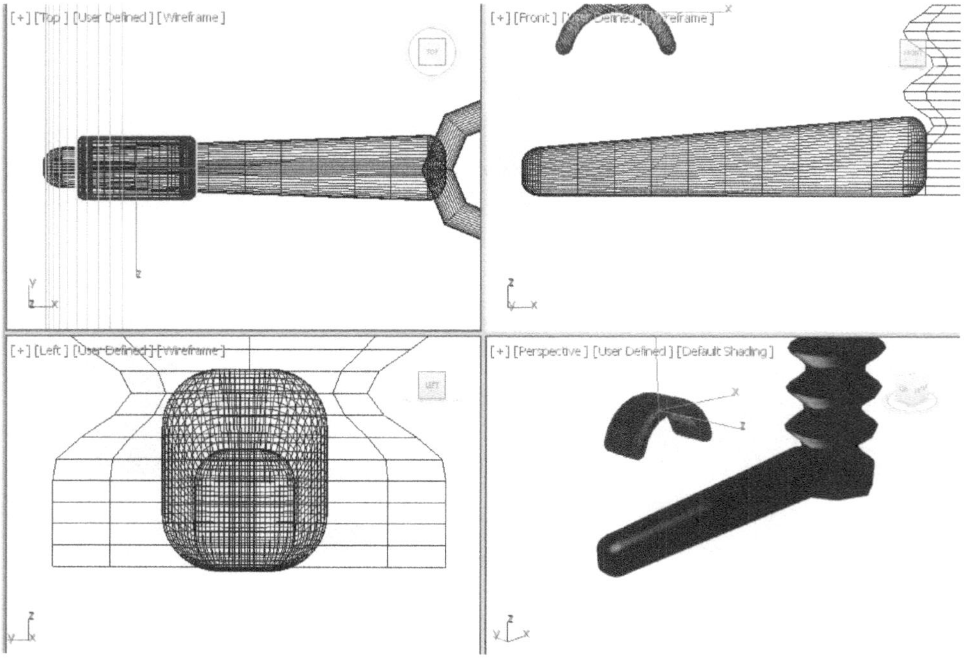

***Figure 3-20** The box01 geometry displayed in viewports after applying the **Bend** modifier*

Next, you need to create a chamfer cylinder.

9. Make sure the **Extended Primitives** option is selected in the drop-down list of the **Command Panel**. Now, choose the **ChamferCyl** tool from the **Object Type** rollout.

10. Activate the Top viewport and expand the **Keyboard Entry** rollout. Set the values as follows:

    X: **-40** Y: **0.0** Z: **-20.0**
    Radius: **5.0** Height: **7.0** Fillet: **1.0**

11. Choose the **Create** button; a chamfer cylinder is created.

12. Make sure the chamfer cylinder is selected. In the **Parameters** rollout, set the values as follows:

    Height Segs: **10** Fillet Segs: **10** Cap Segs: **11**

    Make sure the value in the **Sides** spinner is **12** and the **Smooth** check box is selected.

13. In the **Name and Color** rollout, enter **cylinder01** as the name of the chamfer box and press ENTER. Also, modify its color by specifying the values as follows:
    Red: **151** Green: **151** Blue: **151**

14. Align *cylinder01* with *box01* in the viewports using the **Select and Rotate** and **Select and Move** tools, refer to Figure 3-21.

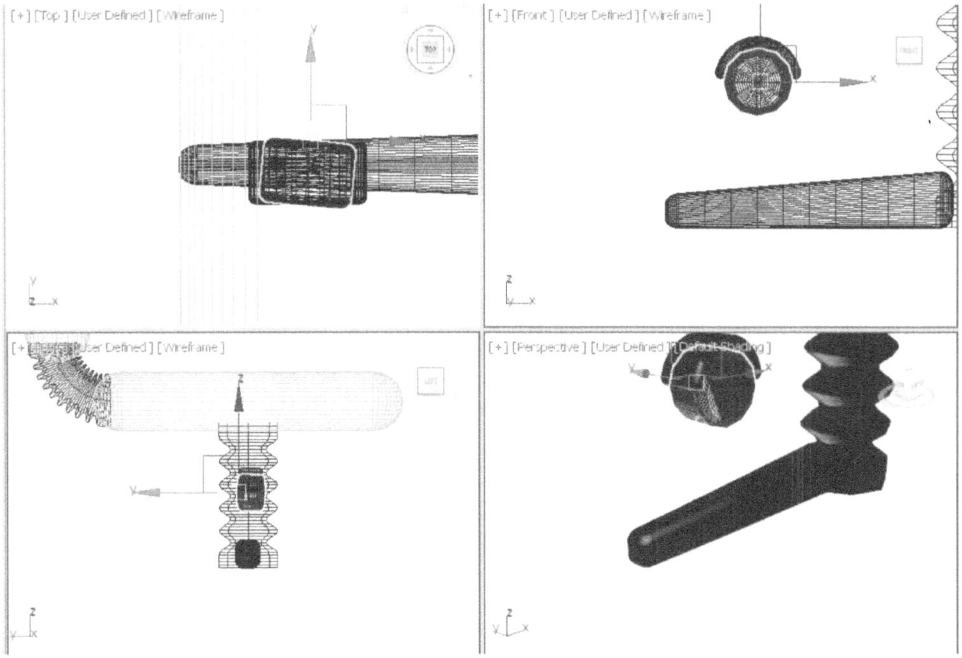

Figure 3-21 *Alignment of cylinder01 with box01 in viewports*

Next, you need to group *cylinder01* and *box01*.

15. Select *cylinder01* and *box01* from the Scene Explorer and right-click; a quad menu is displayed. Choose **Add Selected to > New Group** from the quad menu; the **Group** dialog box is displayed.

16. Enter **roller01** in the **Group name** text box and choose the **OK** button; the group is named as *roller01*.

17. Using the **Select and Move** and **Select and Rotate** tools, align *roller01* with *leg01* in all viewports, as shown in Figure 3-22.

18. Group *leg01* and *roller01* as *f_leg01* as done earlier.

19. Choose the **Zoom Extents All** tool to view all the objects in the viewports.

Next, you need to create other legs for the chair using the same dimensions. To do so, you need to copy *f_leg01*.

20. Activate the Top viewport and make sure *f_leg01* is selected. Now, choose the **Hierarchy** tab from the **Command Panel.** In the **Pivot** tab, choose the **Use Working Pivot** button in the **Working Pivot** rollout; a gizmo is displayed at the center of the scene.

Also, the **USE WP** text is displayed below the viewport label in the viewports.

The working pivot point enables you to rotate *f_leg01* about an arbitrary and persistent point in the scene without affecting the original pivot point of the object.

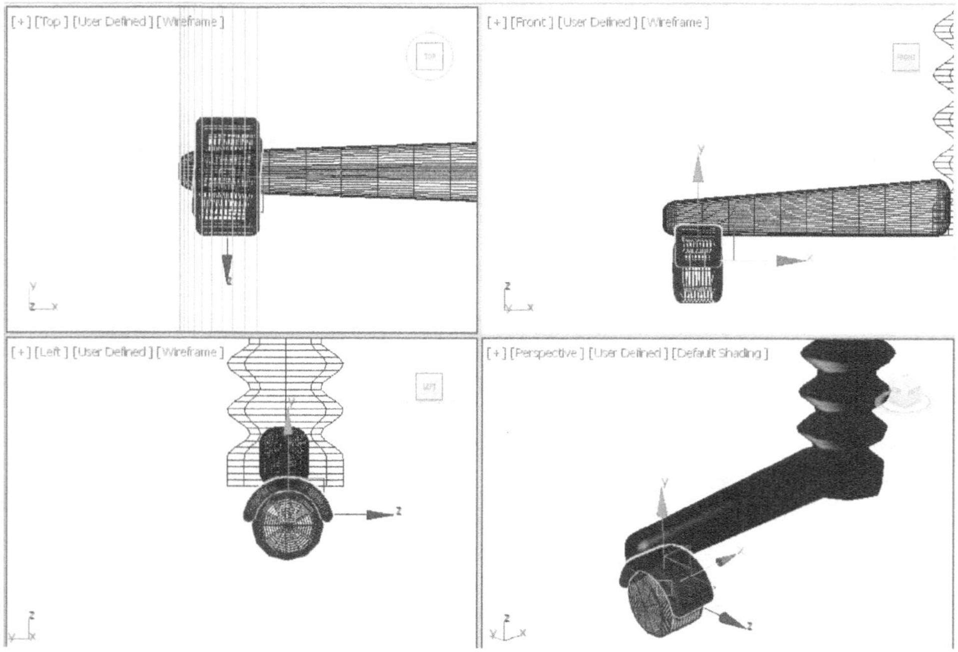

Figure 3-22 *Alignment of roller01 with leg01 in viewports*

21. Activate the Top viewport and then choose the **Select and Rotate** tool; a circular gizmo is displayed, as shown in Figure 3-23.

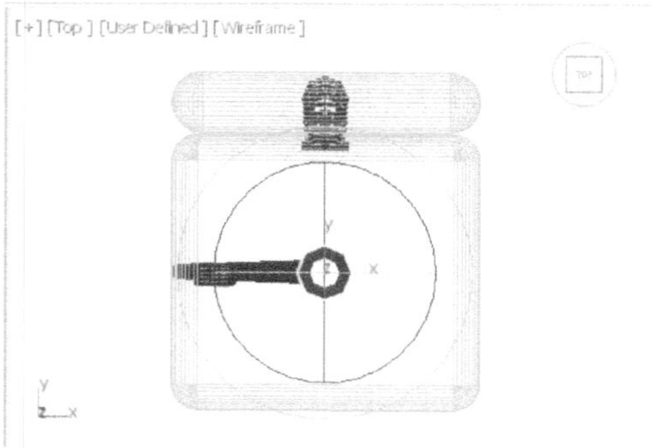

Figure 3-23 *A circular gizmo displayed at the location of the working pivot point*

Figure 3-24 *The f_leg01 group rotated about* ***-72*** *degrees with respect to the working pivot point*

22. Move the cursor over the Z-axis that is blue in color; it get highlighted in yellow color. Now, press and hold the SHIFT key as well as the left mouse button and drag the cursor until the value in the **Z** spinner in the coordinate display becomes about **-72**, refer to Figure 3-24. Release the left mouse button; the **Clone Options** dialog box is displayed. Make sure that the **Copy** radio button is selected. Set **4** in the **Number of Copies** spinner. Choose the **OK** button; *f_leg002*, *f_leg003*, *f_leg004*, and *f_leg005* are displayed in the viewports, as shown in Figure 3-25.

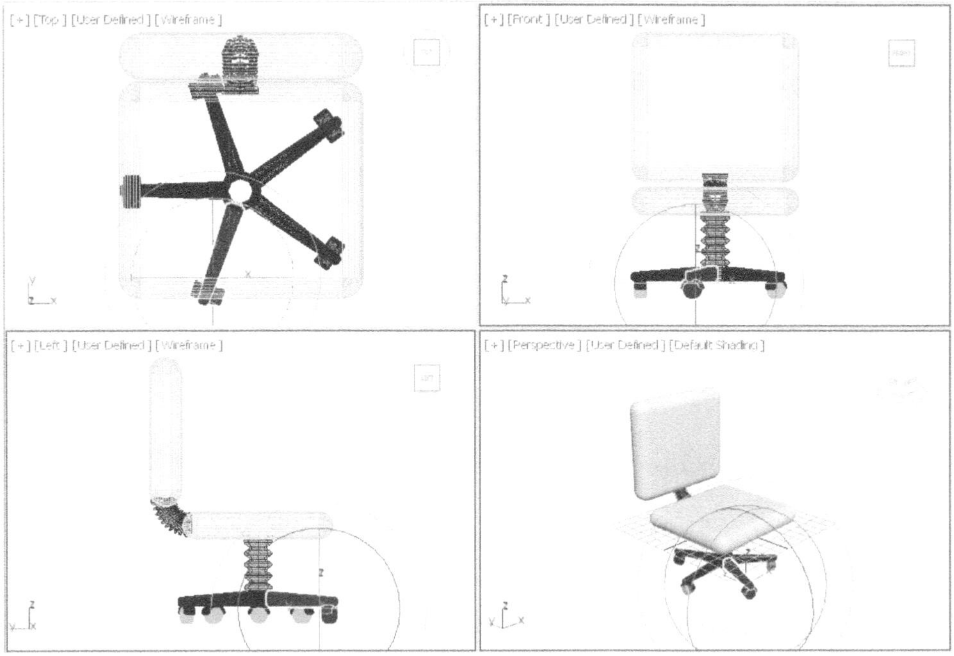

Figure 3-25 The f_leg002, f_leg003, f_leg004, and f_leg005 in viewports

23. Deactivate the **Use Working Pivot** button by choosing it again in the **Working Pivot** rollout; the cursor moves to its original position. Also, the USE WP text disappears from the viewport label in the viewports.

Creating Right Hand Support

In this section, you will use the **Box** tool from **Standard Primitives** to create right hand support of the chair.

1. Activate the Top viewport and choose **Create > Geometry** from the **Command Panel**. Now, select the **Standard Primitives** option from the drop-down list. Also, choose the **Box** tool from the **Object Type** rollout.

2. Expand the **Keyboard Entry** rollout and set the values as follows:

   Length: **5.515** Width: **9.0** Height: **55.0**

3. Choose the **Create** button; a box is created.

4. In the **Parameters** rollout, set the values as given below:

   Length Segs: **15** Width Segs: **15** Height Segs: **15**

Note

The more the number of segments in an object, the smoother will be its surface when you apply a modifier to it.

5. In the **Name and Color** rollout, enter **right hand support01** as the name of the box and press ENTER. Also, change its color to black.

   Next, you need to apply the **Bend** modifier to *right hand support01* to give it the shape of a hand rest.

6. Make sure that *right hand support01* is selected and then apply the **Bend** modifier to it as discussed earlier.

7. In the **Parameters** rollout of the **Bend** modifier, set the parameters as follows:

   Bend Area
   Angle: **107.5** Direction: **270.0**
   Bend Axis Area

   Make sure the **Z** radio button is selected.

8. Align *right hand support01* in all viewports, as shown in Figure 3-26.

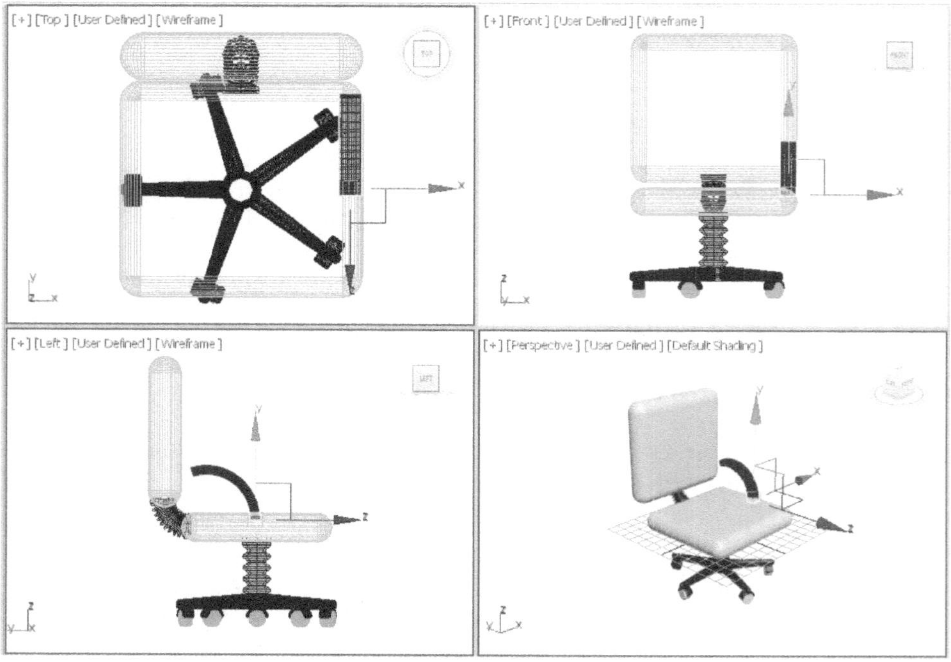

Figure 3-26 *Alignment of right hand support01 in viewports*

Next, you need to create the second part of *right hand support01*. To do so, you need to copy *right hand support01*.

9. Activate the Top viewport, select *right hand support01,* and then create its copy; the copy is automatically named as *right hand support002*.

10. In the **Parameters** rollout of the **Bend** modifier, set the values as follows:

Bend Area
Angle: **90** Direction: **180**

Bend Axis Area
Make sure the **Z** radio button is selected.

Limits Area
Select the **Limit Effect** check box.
Upper Limit: **18.0**

11. After entering the values, *right hand support002* is modified, as shown in Figure 3-27.

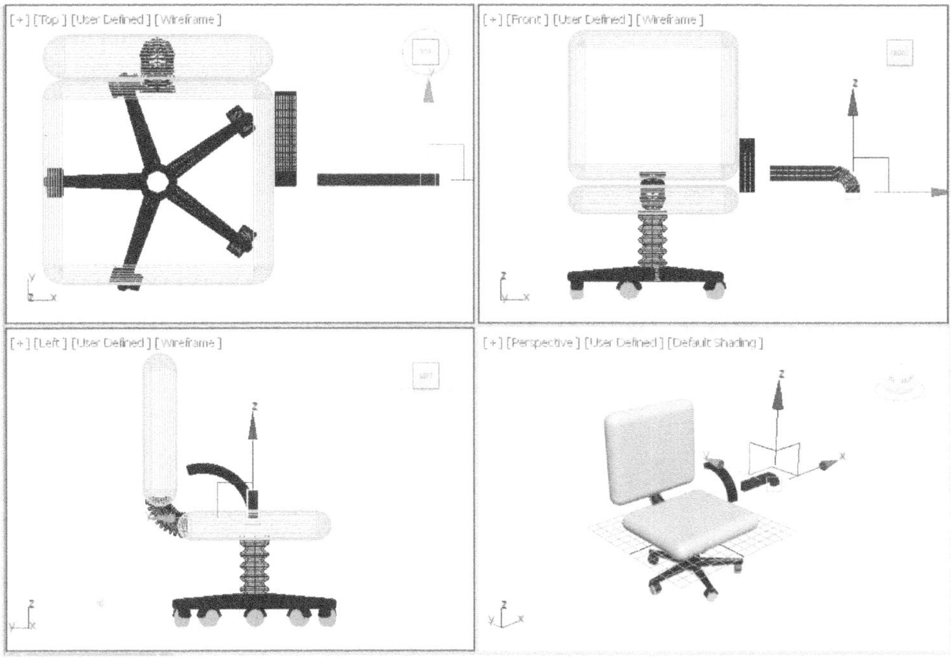

Figure 3-27 *The right hand support002 geometry after modifying the values of the* ***Bend*** *modifier*

12. Align *right hand support002* in viewports using the **Select and Move** and **Select and Rotate** tools, as shown in Figure 3-28.

Note
You can use the ***Orbit*** *tool in the Perspective viewport to view the proper alignment of right hand support002 in viewports.*

13. Select *right hand support01* and *right hand support002* and group them as *right hand support*.

14. Choose the **Zoom Extents All** tool to view all the objects in the viewports.

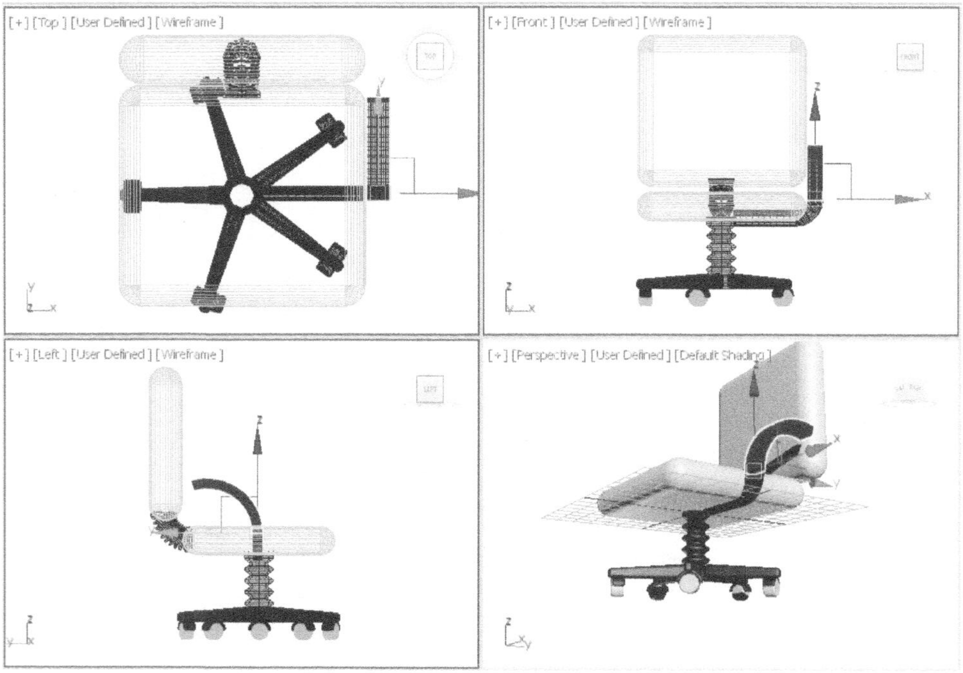

Figure 3-28 Alignment of right hand support002 in viewports

Creating Left Hand Support

In this section, you will create a copy of *right hand support* to create a support for the left hand.

1. Activate the Top viewport, select *right hand support*.

2. Choose the **Mirror** tool from the **Main Toolbar**; the **Mirror: Screen Coordinates** dialog box is displayed, as shown in Figure 3-29.

3. In this dialog box, make sure that the **X** radio button is selected in the **Mirror Axis** area and set -81.4 in the **Offset** spinner. Next, select the **Copy** radio button in the **Clone Selection** area and then choose the **OK** button.

Saving and Rendering the Scene

In this section, you will save and then render the scene. You can also view the final rendered image of this scene by downloading the *c03_3dsmax_2017_rndr.zip* file from *www.cadsofttech.com. www. www.cadsofttech.com*.The path of the file is as follows: *Textbooks > Animation and Visual Effects > 3ds Max > Autodesk 3ds Max 2017 for Novices*.

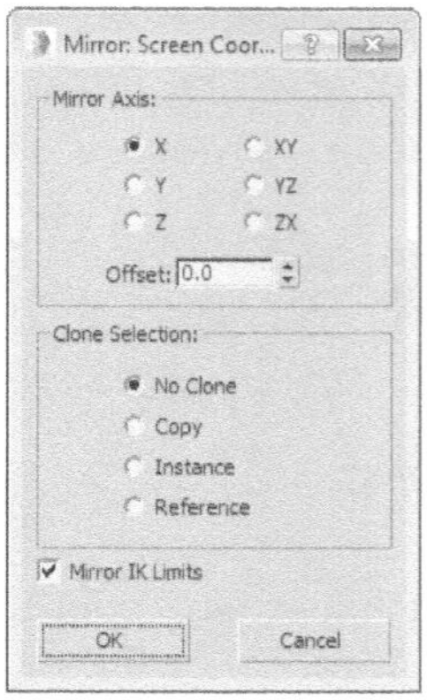

***Figure 3-29** The **Mirror: Screen Coordinates** dialog box*

1. Change the background color of the scene to white by following the steps given in Tutorial 1 of Chapter 2 .

2. Choose **Save** from the **Application** menu.

3. Activate the Perspective viewport. Next, choose the **Render Production** tool from the **Main Toolbar**; the **Rendered Frame** window is displayed showing the final output of the scene, refer to Figure 3-30.

Figure 3-30 *The final rendered output*

EXERCISES

The rendered output of the models used in the following exercises can be accessed by downloading the *c03_3dsmax_2017_exr.zip* from *www.cadsofttech.com*. The path of the file is as follows: Textbooks > Animation and Visual Effects > 3ds Max > Autodesk 3ds Max 2017 for Novices.

Exercise 1

Create the model of a stool shown in Figure 3-31 using your own dimensions.

(Expected time: 30 min)

Figure 3-31 *The model of a stool*

Exercise 2

Create the models shown in Figures 3-32 and 3-33 by using your own dimensions.

(Expected time: 30 min)

Figure 3-32 *The model of a table*

Figure 3-33 *The model of a chair*

Chapter 4

Working with Architectural Objects

Learning Objectives

After completing this chapter, you will be able to:

- *Use the Mirror and Align tools*
- *Create AEC extended objects*
- *Create doors and windows*

INTRODUCTION

In this chapter, you will learn to create the default objects and patch grids in 3ds Max. Also, you will learn to use the **Mirror** and **Align** tools while modeling the objects with the help of various primitives and default objects. In addition, you will learn to create doors and windows.

MIRROR TOOL

Menu bar:	Tools > Mirror
Main Toolbar:	Mirror

The **Mirror** tool is used to mirror or clone the selected object about the center of the current coordinate system. You can also move the object while mirroring its orientation. To mirror an object, select it in a viewport and choose the **Mirror** tool from the **Main Toolbar**; the **Mirror** dialog box will be displayed, refer to Figure 4-1. Set the parameters in this dialog box and choose the **OK** button to mirror the objects.

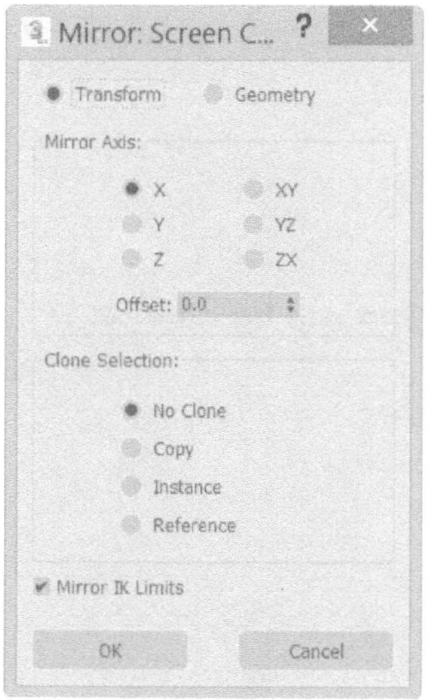

Figure 4-1 *The* ***Mirror*** *dialog box*

ALIGN TOOL

Menu bar:	Tools > Align > Align
Main Toolbar:	Align
Keyboard:	ALT+A

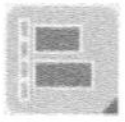

The **Align** tool enables you to align the current object with the target object. To align an object using the **Align** tool, select the current object and then choose the **Align** tool from the **Main Toolbar**; the shape of the cursor will change, as shown in Figure 4-2. Move the cursor over the target object in the viewport and click on it; the **Align Selection (X)** dialog box will be displayed, refer to Figure 4-3. Here, **X** refers to the name of the target object. You need to use the options in the **Align Selection (X)** dialog box to align the objects.

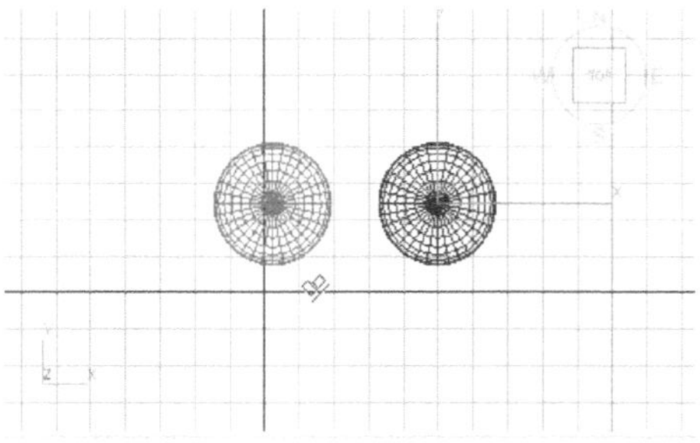

Figure 4-2 The align cursor in the Top viewport

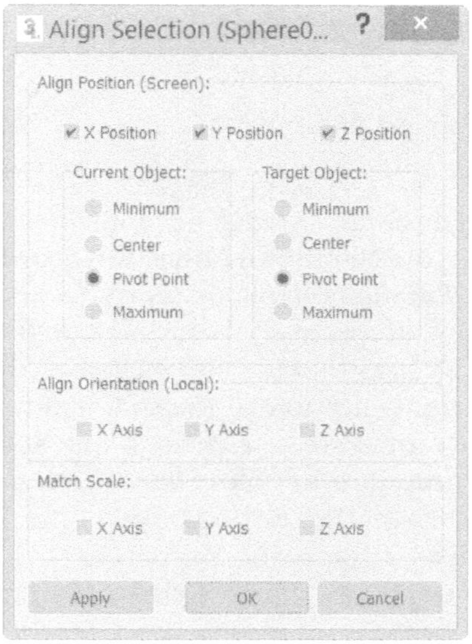

*Figure 4-3 The **Align Selection (Sphere002)** dialog box*

AEC EXTENDED PRIMITIVES

In 3ds Max, there are some in-built objects such as trees, railings, walls, and so on, which are known as AEC extended objects. The AEC extended objects are used in the architectural, engineering, and construction fields. All AEC extended objects can be created dynamically by using the mouse or by entering the parameters in the **Keyboard Entry** rollout.

To create an AEC extended object, you need to choose **Create > Geometry** in the **Command Panel**; the **Standard Primitives** option will be displayed by default in the drop-down list. Select the **AEC Extended** option from the drop-down list and activate the viewport in which you want to create the objects. Next, choose the corresponding tool from the **Object Type** rollout. In this section, you will learn to create and modify the AEC extended objects using various tools available in the **Object Type** rollout.

Creating a Railing

Menu bar: Create > AEC Objects > Railing
Command Panel: Create > Geometry > AEC Extended > Object Type rollout > Railing

To create a railing, activate the viewport in which you want to create it and then choose the **Railing** tool from the **Object Type** rollout; the **Name and Color**, **Railing**, **Posts**, and **Fencing** rollouts will be displayed. Now, press and hold the left mouse button on the left side of a viewport and then drag the cursor to the right side to specify the length of the railing. Release the left mouse button to set the length. Now, move the cursor up to specify the height of the railing and click on the screen; the railing will be created in the viewports. You may need to use the **Zoom Extents All** tool to view the entire railing.

Creating a Wall

Menu bar:	Create > AEC Objects > Wall
Command Panel:	Create > Geometry > AEC Extended > Object Type rollout > Wall

To create a wall, activate the Top viewport and then choose the **Wall** tool from the **Object Type** rollout; the **Name and Color**, **Keyboard Entry**, and **Parameters** rollouts will be displayed. Now, in the **Parameters** rollout, set the width and height by entering values in the **Width** and **Height** spinners, respectively. Next, click on the left of the Top viewport to create the starting point of the wall. Drag the cursor to the right to define the length of the wall and then click on the screen. Now, right-click to exit the command; a wall segment will be created, as shown in Figure 4-4. If you want to create another segment of the wall in continuation, then you need to repeat the same procedure as followed for the first segment without right-clicking. Next, to create a closed wall, click on the starting point of the first wall segment; the **Weld Point?** message box will be displayed. Choose the **Yes** button in this dialog box; a closed wall will be displayed, as shown in Figure 4-5. Next, right-click to exit the command.

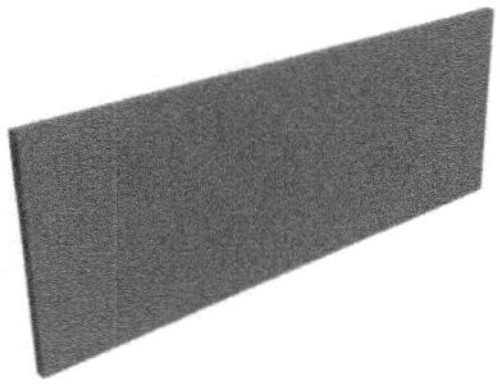

Figure 4-4 *The wall segment displayed*

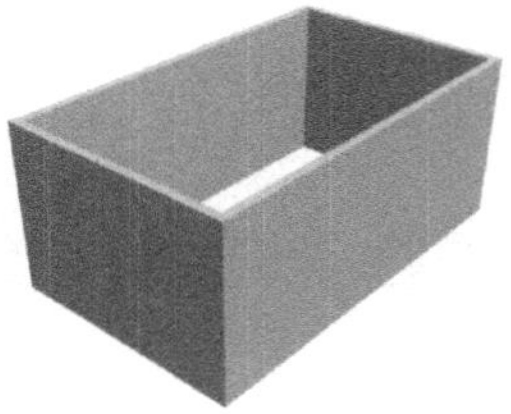

Figure 4-5 *A closed wall displayed*

CREATING DOORS

In 3ds Max, there are three tools to create the in-built doors. These tools are **Pivot**, **Sliding**, and **BiFold**. You can use these doors while creating houses, offices, rooms, and so on. To create doors, you need to choose **Create > Geometry** from the **Command Panel**; the **Standard Primitives** option will be displayed in the drop-down list. Select the **Doors** option from the drop-down list and activate the viewport in which you want to create the doors. Next, choose the corresponding tool from the **Object Type** rollout. In this section, you will learn to create and modify different types of doors using various tools available in the **Object Type** rollout.

Creating a Pivot Door

Menu bar:	Create > AEC Objects > Pivot Door
Command Panel:	Create > Geometry > Doors > Object Type rollout > Pivot

The pivot door is joined or hinged only on one side. To create a pivot door, choose the **Pivot** tool from the **Object Type** rollout; the **Name and Color**, **Creation Method**, **Parameters**, and **Leaf Parameters** rollouts will be displayed.

Activate the Top viewport and press and hold the left mouse button on the left of the viewport. Now, drag the cursor to the right of the viewport and release the left mouse button to define the width. Next, move the cursor up or down to define the depth of the door and click on the screen to set the depth. Now, again move the cursor up or down to define the height of the door. Click on the screen; the pivot door will be created in all viewports, refer to Figure 4-6.

Figure 4-6 *The pivot door created in the viewport*

CREATING WINDOWS

In 3ds Max, there are six tools to create different types of default windows such as **Awning**, **Casement**, and so on. You can use these windows at various places for architectural designs.

To use these tools for creating the windows, choose **Create > Geometry** in the **Command Panel**; the **Standard Primitives** option will be displayed in the drop-down list. Now, select the **Windows** option from this drop-down list; various tools will be displayed in the **Object Type** rollout. In this section, you will learn to create various types of windows using these tools.

Creating an Awning Window

Menu:	Create > AEC Objects > Awning Window
Command Panel:	Create > Geometry > Windows > Object Type rollout > Awning

An awning window has one or more cases that are joined at its top. To create an awning window, choose the **Awning** tool from the **Object Type** rollout; the **Name and Color**, **Creation Method**, and **Parameters** rollouts will be displayed.

Activate the Top viewport and press and hold the left mouse button on the left side of the viewport, drag the cursor to the right side to specify the width of the window, and release the left mouse button. Next, move the cursor up or down to define the depth of the window and click on the screen. Now, move the cursor up or down to specify the height of the window. Click on the screen; the awning window will be created in all viewports, refer to Figure 4-7. The options in the **Name and Color** and

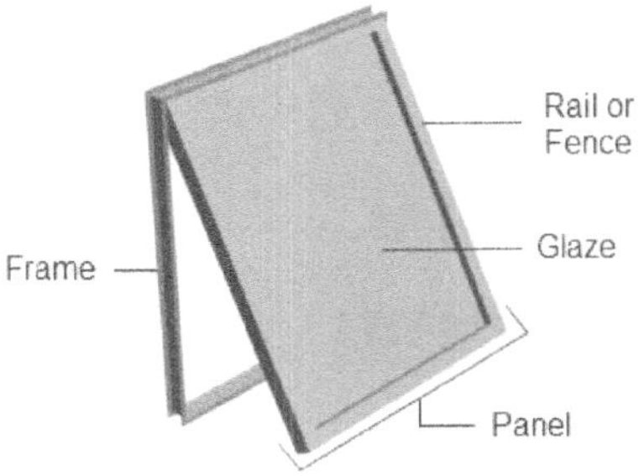

Figure 4-7 *An awning window*

CREATING STAIRS

In 3ds Max, there are four tools that are used to create different types of default stairs such as **LType Stair**, **Spiral Stair**, and so on. To invoke these tools, choose **Create > Geometry** in the **Command Panel**; the **Standard Primitives** option will be displayed in the drop-down list. Now, select the **Stairs** options from the drop-down list; various tools will be displayed in the **Object Type** rollout. In this section, you will learn to create various types of stairs using these tools.

Creating L-Type Stairs

Menu bar:	Create > AEC Objects > L-Type Stair
Command Panel:	Create > Geometry > Stairs > Object Type rollout > L-TypeStair

The L-type stairs have two stairways that are joined at right angles to each other. To create the L-type stairs, choose the **L-Type Stair** tool from the **Object Type** rollout; the **Name and Color**, **Parameters**, **Carriage**, **Railings**, and **Stringers** rollouts will be displayed.

Now, activate the Top viewport. Press and hold the left mouse button on the left side of the viewport, drag the cursor to the right side of the viewport to specify the length of the first stairway of the stairs, and then release the left mouse button. Next, move the cursor up or down to define the length of the second stairway of the stairs and then click on the screen. Now, move the cursor up to specify the overall height of the stair and then click on the screen; the L-type stairs will be created in the viewport, as shown in Figure 4-8.

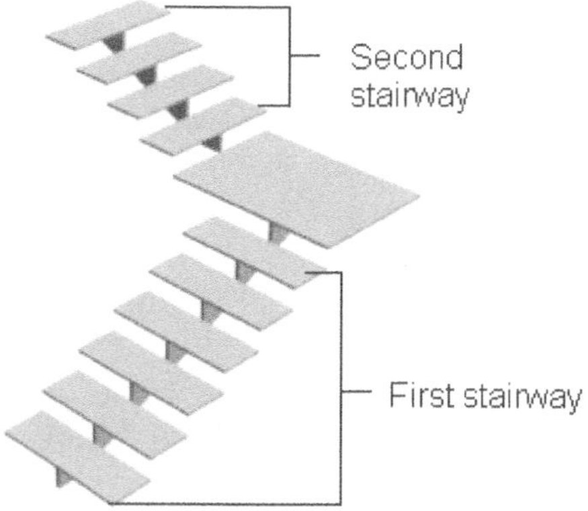

Figure 4-8 *The L-type open stairs created*

TUTORIAL

Tutorial 1

In this tutorial, you will create a nature scene, as shown in Figure 4-9, using the AEC extended objects and standard primitives. **(Expected time: 30 min)**

The following steps are required to complete this tutorial:

a. Create the project folder.
b. Create floor.
c. Create railings.
d. Create trees.
e. Create gate.
f. Save and render the scene.

Figure 4-9 *The nature scene*

Creating the Project Folder

Create a new project folder with the name *c04_tut1* at *\Documents\3dsmax2017* and then save the scene with the name *c04tut1*, as discussed in Tutorial 1 of Chapter 2.

Creating Floor

In this section, you will create a floor using the **Plane** tool.

1. Activate the Top viewport. Choose **Create > Geometry** in the **Command Panel**; the **Standard Primitives** option is displayed in the drop-down list. Choose the **Plane** tool from the **Object Type** rollout.

2. In the **Keyboard Entry** rollout, set the parameters as follows:

   Length: **1485.0** Width: **1605.15**

3. Choose the **Create** button from the **Keyboard Entry** rollout; a plane is displayed in viewports. Now, choose the **Zoom Extents All** tool to display the entire plane in the viewports.

4. In the **Name and Color** rollout, enter **floor**; the plane is named as *floor*.

5. Use the color swatch in the **Name and Color** rollout to modify the color of *floor* to green.

6. Activate the Perspective viewport and set the view using the **Zoom** and **Orbit** tools, as shown in Figure 4-10.

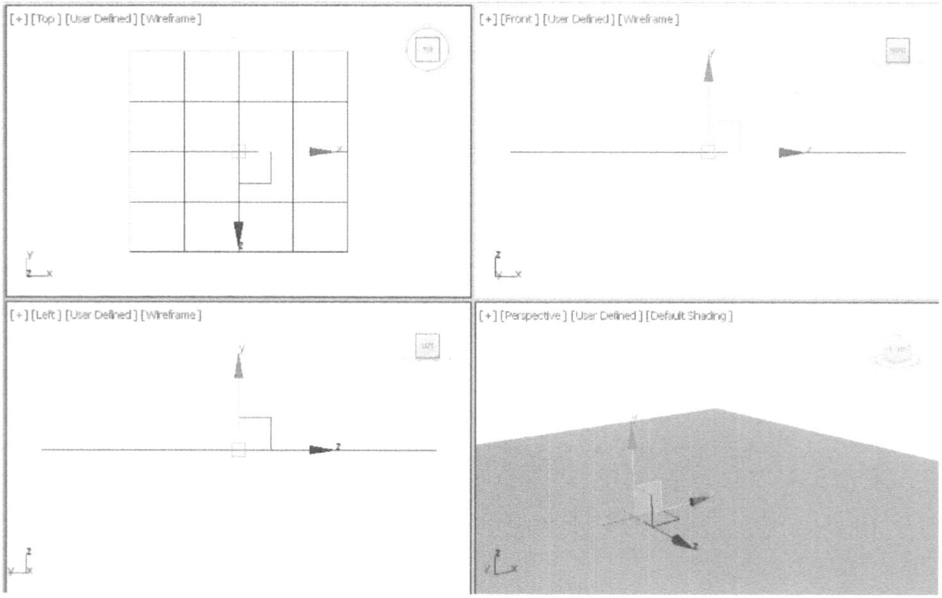

Figure 4-10 *The floor geometry after using the* ***Zoom*** *and* ***Orbit*** *tools*

Creating Railings

In this section, you will create railings around *floor* using the **Railing** tool.

1. Choose **Create > Geometry** in the **Command Panel**; a drop-down list is displayed below the **Geometry** button. Select the **AEC Extended** option from the drop-down list and then choose the **Railing** tool from the **Object Type** rollout.

2. Activate the Top viewport. Next, move the cursor to the upper left corner of *floor*, press and hold the left mouse button, and drag the cursor to the upper right corner of *floor* to specify the length of the railing. Release the left mouse button to set the length, and then move the cursor up to specify the height of the railing. Click on the viewport; a railing is created.

3. In the **Name and Color** rollout, enter **railing01** and press ENTER.

4. Use the color swatch to modify the color of *railing01* and enter the values as follows:

   Red: **177** Green: **88** Blue: **26**

5. Make sure *railing01* is still selected. Choose the **Modify** tab in the **Command Panel**; the **Railing**, **Posts**, and **Fencing** rollouts are displayed.

6. In the **Railing** rollout, set the parameters as follows:

   Length: **1400.0**

   Top Rail area
   Profile: **Round** Depth: **10.0**
   Width: **3.0** Height: **100.0**

   Lower Rail(s) area
   Profile: **Round** Depth: **4.0**
   Width: **3.0**

7. Choose the **Lower Rail Spacing** button in the **Lower Rail(s)** area; the **Lower Rail Spacing** dialog box is displayed. Make sure the **Count** check box is selected. Next, set the value **3** in the spinner on the right side of the **Count** check box and choose the **Close** button; the railing is displayed in all viewports, as shown in Figure 4-11.

8. In the **Posts** rollout, set the values as follows:

   Profile: **Round** Depth: **7.0**
   Width: **5.0** Extension: **2.0**

9. Choose the **Post Spacing** button in the **Post** rollout; the **Post Spacing** dialog box is displayed. Make sure the **Count** check box is selected. Next, set the value **4** in the spinner on the right side of the **Count** check box and choose the **Close** button.

10. In the **Fencing** rollout, select **(none)** from the **Type** drop-down list.

Note

If you want to add fencing to the railing, then select the type of fencing from the ***Type*** *drop-down list and set the parameters in the respective area.*

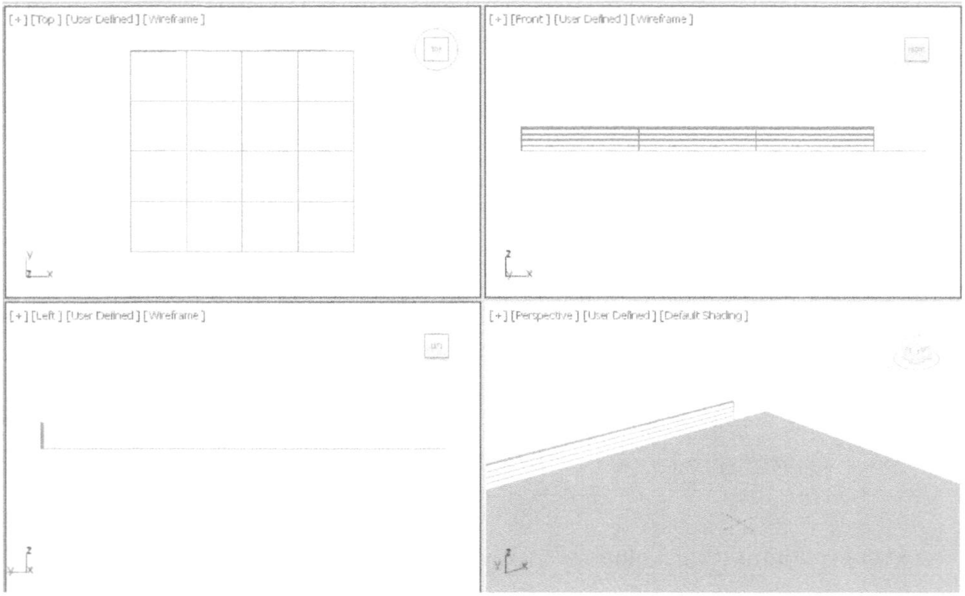

Figure 4-11 *The railing01 displayed in viewports*

11. Make sure *railing01* is selected in the Top viewport. Now, choose the **Select and Move** tool from the **Main Toolbar** and align the railing manually, as shown in Figure 4-12.

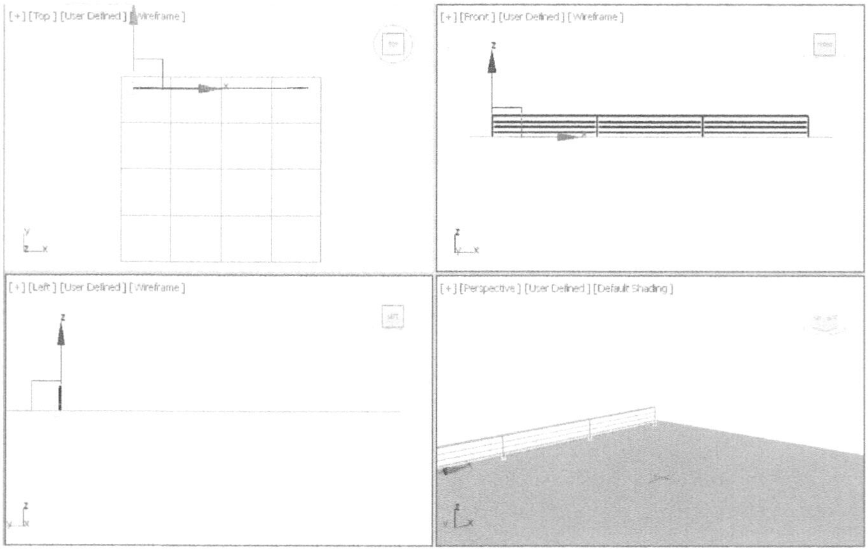

Figure 4-12 *The railing01 geometry in viewports after aligning it in the Top viewport*

Next, you need to create one more railing.

12. Activate the Top viewport and choose the **Railing** tool. Create one more railing at the right angle of *railing01* using the same method as discussed for creating *railing01*, refer to Figure 4-13

Note

When you create another railing, all the dimensions of railing01, except the length and height, will be taken automatically.

13. In the **Name and Color** rollout, enter **railing02**.

14. Assign the same color to *railing02* as was assigned to *railing01*.

15. Make sure *railing02* is selected. Next, choose the **Modify** tab in the **Command Panel**; the **Railing**, **Post**, and **Fencing** rollouts are displayed.

16. In the **Railing** rollout, set the parameters as follows:
    Length: **600.0**

    Top Rail area
    Height: **100.0**

17. Choose the **Select and Move** and **Select and Rotate** tools and align *railing02* in the viewports, as shown in Figure 4-13.

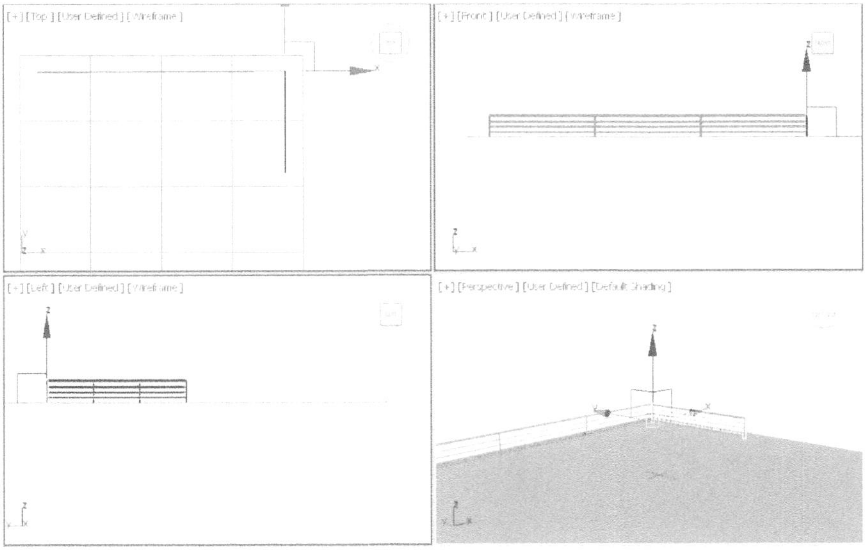

Figure 4-13 *The railing02 geometry in viewports after alignment*

Next, you need to copy *railing02*.

18. Activate the Top viewport. and make sure *railing02* is selected. Next, place the cursor over the vertical axis. Press and hold the SHIFT key and drag *railing02* downward until the value in the **Y** spinner in the Coordinate display becomes around **-601**. Release the left mouse button and the SHIFT key; the **Clone Options** dialog box is displayed.

19. In the **Clone Options** dialog box, make sure the **Copy** radio button is selected. Set the value in the **Number of Copies** spinner to **1** and enter *railing03* in the **Name** text box. Next, choose the **OK** button; *railing 03* is created, as shown in Figure 4-14.

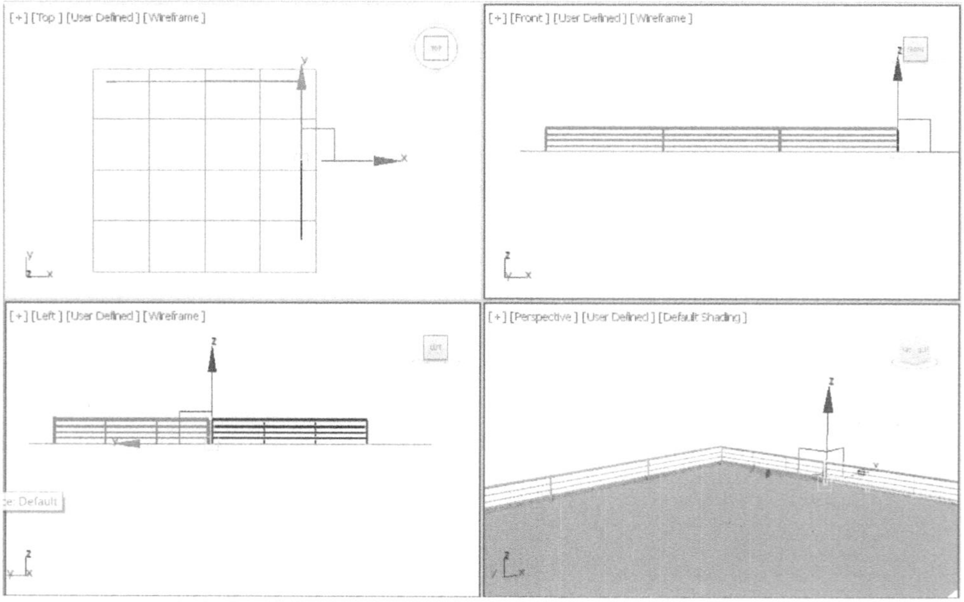

Figure 4-14 *The railing03 geometry after alignment*

20. Activate the Top viewport and select *railing03*. Choose the **Select and Rotate** tool, move the cursor over the Z-axis, and rotate it to -45 degrees in the clockwise direction, as shown in Figure 4-15.

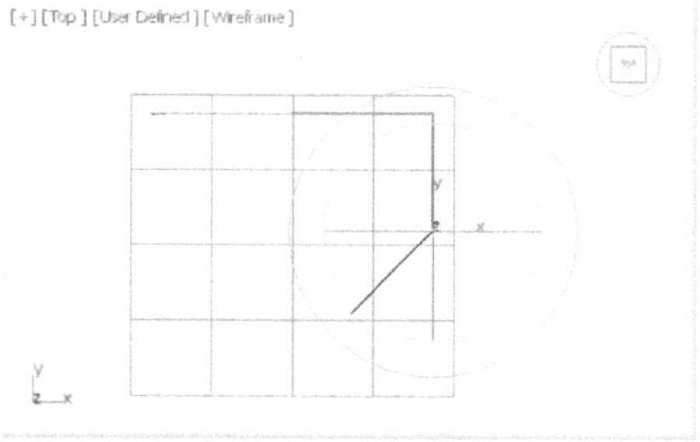

Figure 4-15 *The railing03 geometry rotated in the Top viewport*

Note

*While rotating railing03, choose the **Angle Snap Toggle** tool from the **Main Toolbar** to measure the angle of rotation in increments.*

Now, you need to create the fourth railing.

21. Activate the Top viewport and create another railing starting from the endpoint of *railing03*.

22. In the **Name and Color** rollout, enter **railing04**.

23. Assign the same color to *railing04* as was assigned to *railing01*.

24. Make sure *railing04* is selected. Choose the **Modify** tab in the **Command Panel**; the **Railing**, **Post**, and **Fencing** rollouts are displayed.

25. In the **Railing** rollout, set the parameters as follows:

    Length: **325.0**
    Top Rail area
    Height: **100.0**

26. Make sure *railing04* is aligned in the viewports, as shown in Figure 4-16.

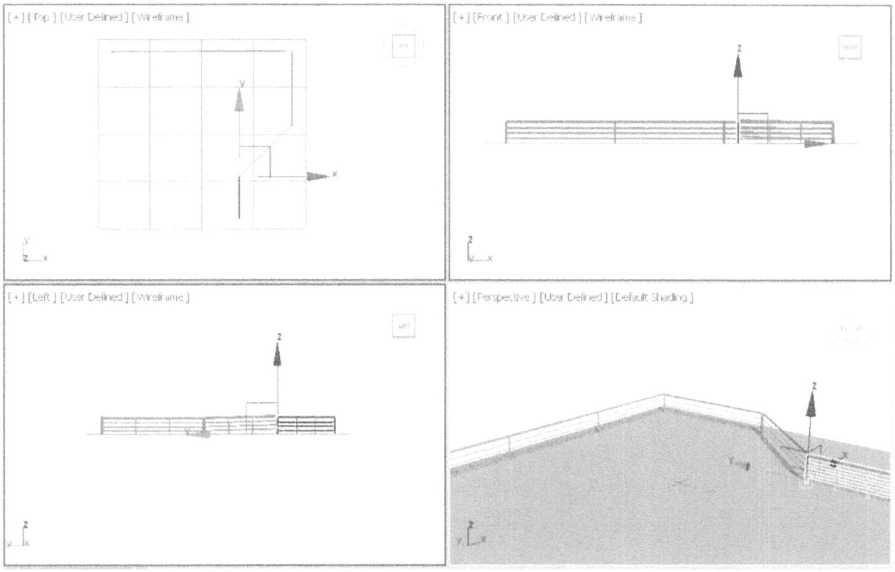

Figure 4-16 *The railing04 geometry aligned in viewports*

27. Select *railing02*, *railing03*, and *railing04* from the Scene Explorer; three railings are selected in the viewports.

28. Make sure the Top viewport is activated and then choose the **Mirror** tool from the **Main Toolbar**; the **Mirror: Screen Coordinates** dialog box is displayed. In the **Mirror Axis** area of this dialog box, make sure the **X** radio button is selected. In the **Offset** spinner, set the value **-980**. In the **Clone Selection** area, select the **Copy** radio button and choose the **OK** button; the copy of all the railings created earlier is displayed and they are automatically named as *railing005*, *railing006*, and *railing007*, refer to Figure 4-17.

29. Adjust the view in the Perspective viewport using the tools in the viewport navigation controls, refer to Figure 4-17.

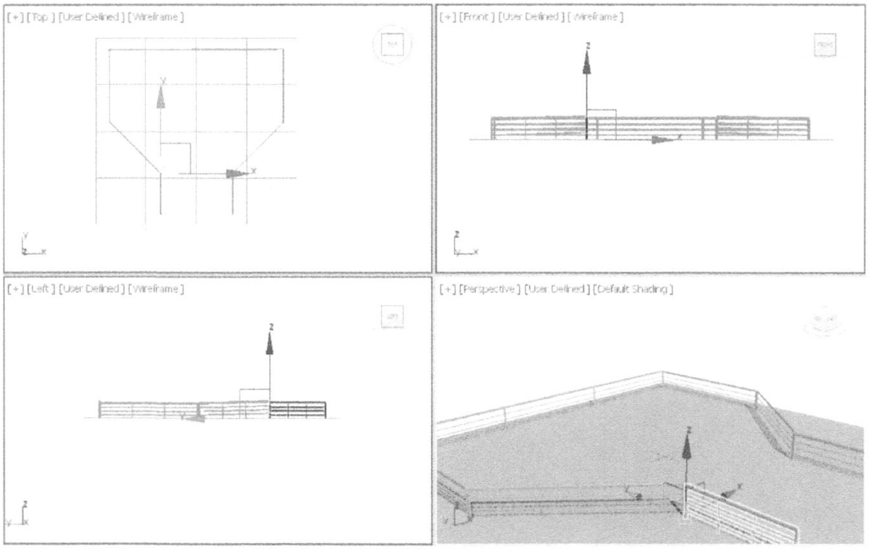

Figure 4-17 The railings geometry displayed in viewports

Creating Trees

In this section, you will create trees using the **Foliage** tool.

1. Choose **Create > Geometry** in the **Command Panel**; a drop-down list is displayed below the **Geometry** button. Select the **AEC Extended** option from the drop-down list and then choose the **Foliage** tool from the **Object Type** rollout.

2. Activate the Top viewport. In the **Favorite Plants** rollout, double-click on **Generic Palm**; the tree is created and displayed in all viewports.

Note

*If the **Generic Palm** tree is not displayed in the palette of the **Favorite Plants** rollout, then choose the **Plant Library** button at the bottom of the rollout; the **Configure Palette** dialog box is displayed. Double-click on the **Generic Palm** tree; the **yes** option is displayed in the **Fav.** column indicating that the tree will be available in the palette. Next, choose the **OK** button.*

3. In the **Name and Color** rollout, enter **tree01**; the tree is named as *tree01*.

4. Make sure *tree01* is still selected. Choose the **Modify** tab in the **Command Panel**; the **Parameters** rollout is displayed.

5. In the **Parameters** rollout, set the following parameters:

   Height: **300.0**

   Use the default values for other options.

6. In the Top viewport, choose the Select and Move tool and move tree01 toward railing01 in all viewports, as shown in Figure 4-18.

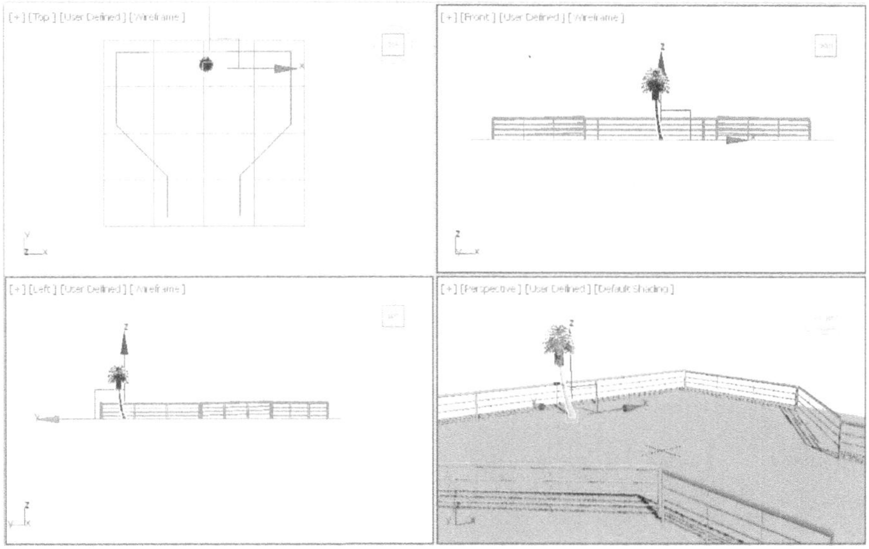

Figure 4-18 *Alignment of tree01 in viewports*

7. Create multiple copies of *tree01* in the Top viewport and align them using the **Select and Move** tool, as shown in Figure 4-19.

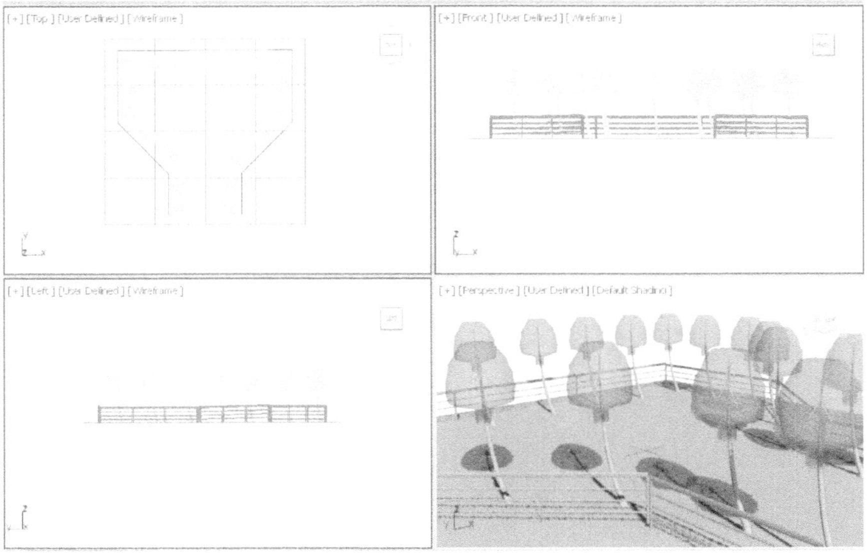

Figure 4-19 *Multiple copies of tree01 created and aligned in viewports*

Creating the Gate

In this section, you will create the gate using the **Pivot** and the **Cylinder** tools.

1. Choose **Create > Geometry** in the **Command Panel**; a drop-down list is displayed below the **Geometry** button. Select the **Doors** option from the drop-down list and then choose the **Pivot** tool from the **Object Type** rollout.

2. Activate the Top viewport and press and hold the left mouse button on the left of the viewport. Now, drag the cursor to the right of the viewport and release the left mouse button to define the width. Next, move the cursor up to define the depth of the door and click on the screen to set the depth. Now, again move the cursor up to define the height of the door. Click on the screen; the pivot door is created in all viewports.

3. In the **Name and Color** rollout, enter **gate**.

4. Use the color swatch to modify the color of *gate* and enter the values as follows:

   Red: **49** Green: **13** Blue: **6**

5. In the **Parameters** rollout, set the following parameters:

   Height: **92** Width: **532** Depth: 9.234
   Open: **20**

   Also, select the **Double Doors** and **Flip Swing** check boxes.

6. In the **Leaf Parameters** rollout, set the following parameters:

   Thickness: **7.7** Stiles/Top Rail: **6.48** Bottom rail: **13.44**
   #Panels Horiz.: **6** #Panels Vert.: **4**

7. In the **Panels** area, select the **Beveled** radio button and then set the following parameters:

   Thickness 1: **0.74** Thickness 2: **1.96** Middle Thick.: **0.991**
   Width 1: **4.871** Width 2: **1.364**

8. Align *gate* using the **Select and Move** and **Select and Rotate** tools, as shown in Figure 4-20. Make sure that there is equal space at both sides of *gate* to place the rods to be created later.

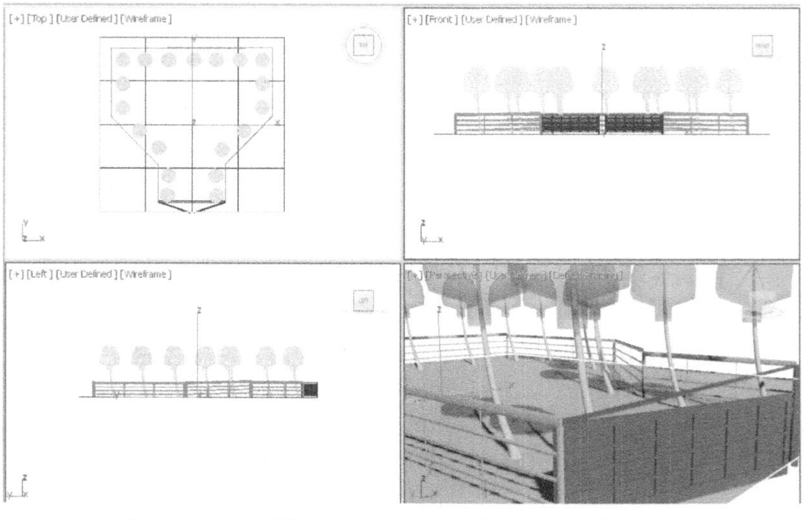

Figure 4-20 *The gate geometry aligned in viewports*

9. In the **Parameters** rollout, set **0** in the **Open** spinner to close *gate*.

   Next, you will create rods at both sides of *gate*.

10. Activate the Top viewport. Make sure the **Geometry** button is chosen in the **Command Panel**. Next, select **Standard Primitives** from the drop-down list located below it. Now, choose the **Cylinder** tool from the **Object Type** rollout.

11. In the **Keyboard Entry** rollout, set the parameters as follows:

    Radius: **2.4** Height: **94**

    Choose the **Create** button; a cylinder is created.

12. In the **Name and Color** rollout, enter **rod1**. Also, assign black color to it.

13. Align *rod1* in the viewports, as shown in Figure 4-21.

14. Create a copy of *rod1* as done earlier and align it, as shown in Figure 4-22.

Next, you will create handles for *gate*.

15. Activate the Top viewport. Choose **Create > Geometry** in the **Command Panel**. Make sure that the **Standard Primitives** option is selected in the drop-down list and then choose the **Cylinder** tool from the **Object Type** rollout.

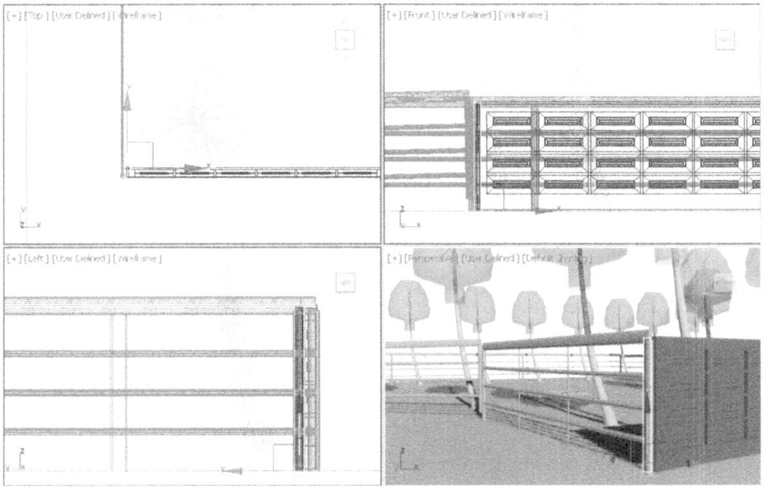

Figure 4-21 *The rod1 geometry aligned in viewports*

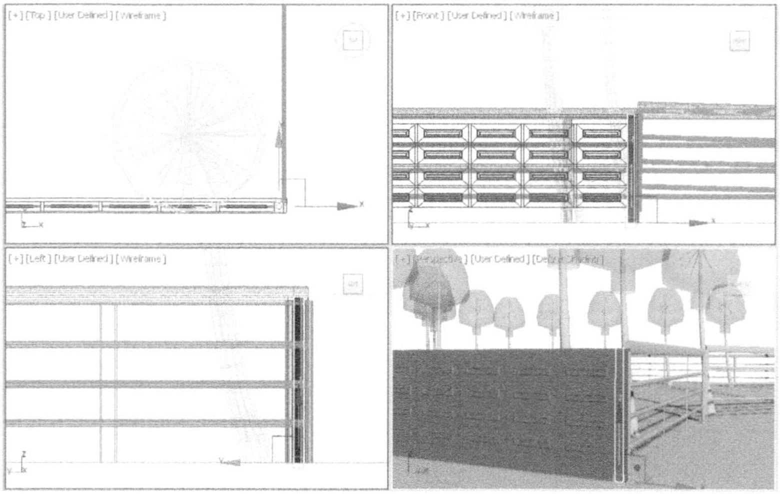

Figure 4-22 *The rod2 geometry aligned in viewports*

16. In the **Keyboard Entry** rollout, set the parameters as follows:

    Radius: **1.5** Height: **35**

    Choose the **Create** button; a cylinder is created.

17. In the **Name and Color** rollout, enter **handle1** and press ENTER. Also, assign black color to it.

18. Make sure *handle1* is selected. In the **Parameters** rollout, set **15** in the **Height Segments** spinner.

    Next, you will apply the **Bend** modifier to *handle1*.

19. Make sure *handle1* is selected and then choose **Modifiers > Parametric Deformers > Bend** from the menu bar; the **Bend** modifier is displayed in the modifier stack. Also, the **Parameters** rollout is displayed in the **Modify** tab.

20. In the **Parameters** rollout of the **Bend** modifier, set the values as follows:

    Bend Area
    Angle: **180** Direction: **270**

    Bend Axis area
    Make sure the **Z** radio button is selected.

21. Activate the Perspective viewport. Next, align *handle1* using the **Select and Place** tool, as shown in Figure 4-23.

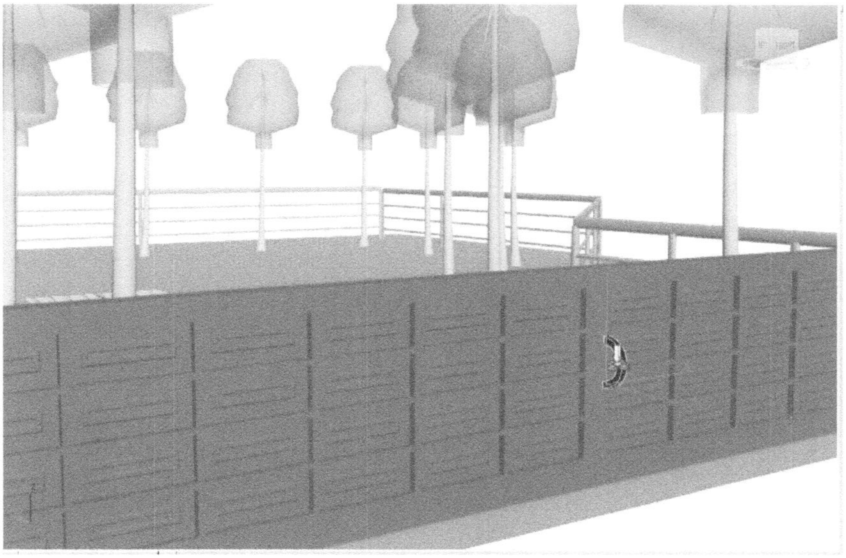

Figure 4-23 *The handle1 geometry aligned*

22. Activate the Front viewport and create a copy of *handle1* and align it, as shown in Figure 4-24.

Saving and Rendering the Scene

In this section, you will save the scene and then render it. You can also view the final rendered image of this scene by downloading the *c04_3dsmax_2017_rndr.zip* file from *www.www.cadsofttech.com*.The path of the file is as follows: *Textbooks > Animation and Visual Effects > 3ds Max > Autodesk 3ds Max 2017 for Novices.*

1. Change the background color of the scene to light blue, as discussed in Tutorial 1 of Chapter 2, using the following parameters:

   Red: **145** Green: **241** Blue: **244**

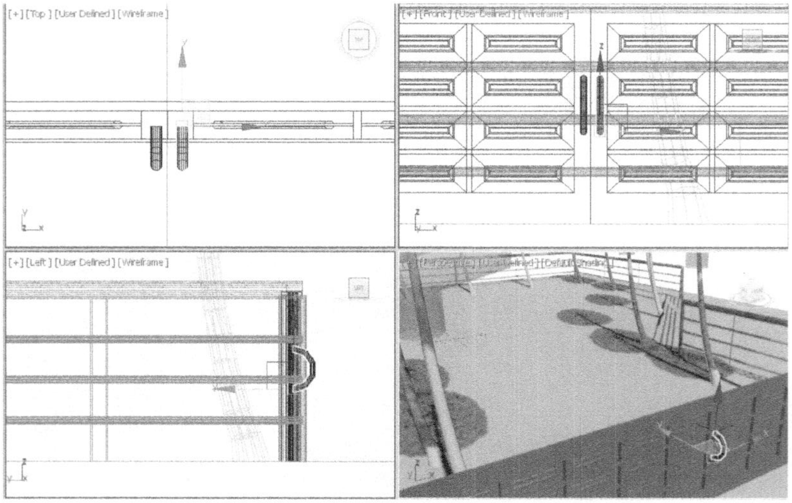

Figure 4-24 The copy of handle1 aligned

2. Choose **Save** from the **Application** menu.

3. Set the render view the Perspective viewport and choose the **Render Production** tool from the **Main Toolbar**; the **Rendered Frame** window is displayed. This window shows the final output of the scene, refer to Figure 4-25.

Figure 4-25 The final output after rendering

EXERCISE

The rendered output of the models used in the following exercise can be accessed by downloading the *c04_3dsmax_2017_exr.zip* from *www.cadsofttech.com*. The path of the file is as follows: Textbooks > Animation and Visual Effects > 3ds Max > Autodesk 3ds Max 2017 for Novices.

Exercise 1

Create the model shown in Figure 4-26 using the dimensions of your choice.

(Expected time: 15 min)

Figure 4-26 *The model for Exercise 1*

This page is intentionally left blank

Chapter 5

Splines and Extended Splines

Learning Objectives

After completing this chapter, you will be able to:

- *Use the AutoGrid option*
- *Create splines*

INTRODUCTION

3ds Max has several basic 2D geometric shapes known as splines and extended splines. They are used to create 3D objects, 3D surfaces, and paths for lofting and creating animation. All splines and extended splines can be created dynamically by using the mouse or by entering the parameters in the **Keyboard Entry** rollout. In this chapter, you will learn about splines and extended splines.

CREATING SPLINES

To create splines, choose **Create > Shapes** in the **Command Panel**; the **Splines** option will be displayed in the drop-down list below the **Shapes** button. Now, activate the viewport in which you want to create the spline by clicking on it and then choose the corresponding tool from the **Object Type** rollout. In this section, you will learn to create and modify the splines and extended splines using the various tools available in the **Object Type** rollout.

Note

*By default, the **Start New Shape** check box is selected in the **Object Type** rollout. This check box enables you to create new shapes separately in the viewport using different tools such as **Line**, **Circle**, **Arc**, and so on available in the **Object Type** rollout. If you clear this check box and then create new shapes, the newly created shapes will be added to the current shape in the viewport. However, the **Section** shape is an exception.*

Creating a Line Spline

Menu bar:	Create > Shapes > Line
Command Panel:	Create > Shapes > Splines > Object Type rollout > Line

To create a line spline, activate the viewport and then choose the **Line** tool from the **Object Type** rollout; various rollouts will be displayed, as shown in Figure 5-1.

To create a line spline dynamically, click in the Top viewport to specify the first vertex or the start point of the line. Now, move the cursor to define the distance between the first and the second vertex, and click on the viewport to create the second vertex. Again, move the cursor and click to create additional vertices. Next, right-click to end the command; an open line spline will be created, as shown in Figure 5-2. To create a closed line spline, instead of right-clicking at the end, click on the first vertex; the **Spline** message box will be displayed, as shown in Figure 5-3. Choose the **Yes** button; a closed line spline will be created, as shown in Figure 5-4.

You can also create a line spline using another method. Click in the viewport to specify the first vertex and then move the cursor to define the distance between the first and the second vertex. Now, press and hold the left mouse button and drag the cursor to create additional vertices; a line spline with a bezier vertex and smooth curve will be created. The bezier vertex is controlled by two points on both endpoints, which help in modifying the curve of the line.

While creating a line spline, if you click in the viewport to create additional vertices, they will become corner vertices, as shown in Figure 5-5. But, if you click and drag to create additional vertices, the curve between the vertices will be smooth and the vertices will be bezier, as shown in Figure 5-6, depending on the options selected from the **Drag Type** and **Initial Type** areas of the **Creation Method** rollout.

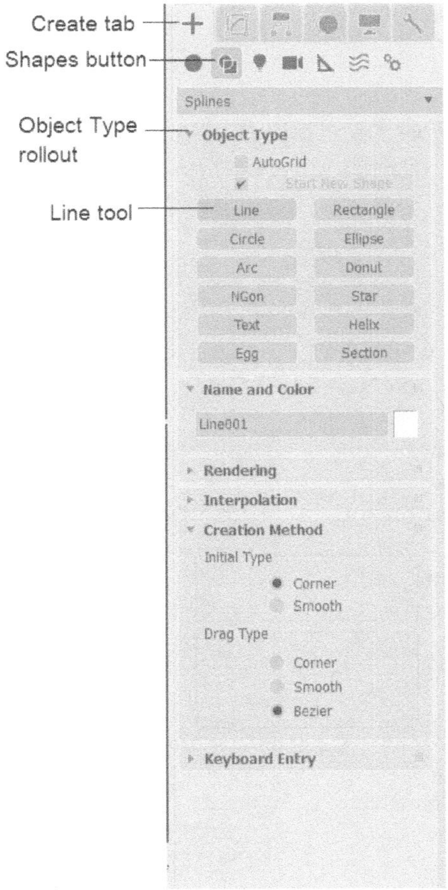

***Figure 5-1** Various rollouts displayed to create and modify a line spline*

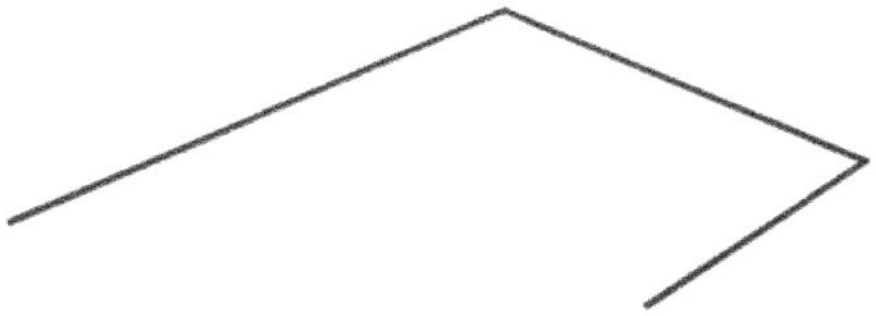

***Figure 5-2** An open line spline created*

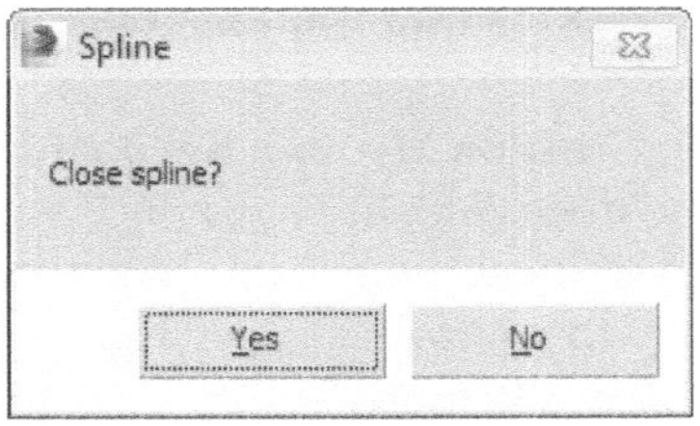

***Figure 5-3** The **Spline** message box*

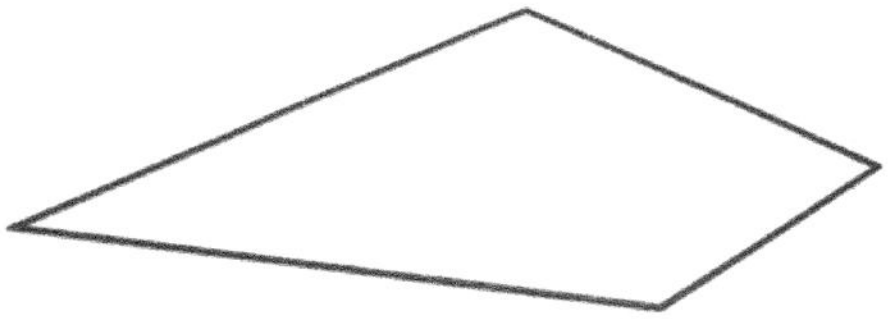

***Figure 5-4** A closed line spline*

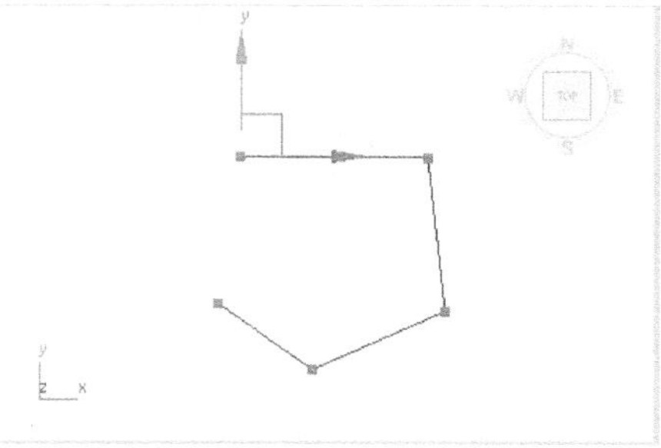

Figure 5-5 A line with corner vertices

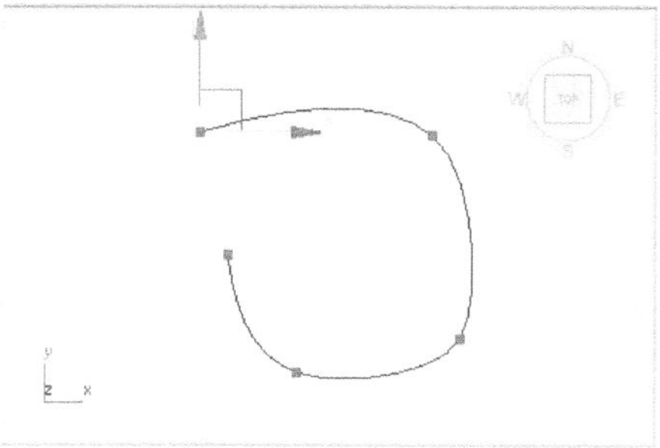

Figure 5-6 A line spline with bezier vertex and smooth curve

Creating a Rectangular Spline

Menu bar:	Create > Shapes > Rectangle
Command Panel:	Create > Shapes > Splines > Object Type rollout > Rectangle

To create a rectangular spline, activate the viewport in which you want to create it and then choose the **Rectangle** tool from the **Object Type** rollout; the **Name and Color**, **Rendering**, **Interpolation**, **Creation Method**, **Keyboard Entry**, and **Parameters** rollouts will be displayed, as shown in Figure 5-7. Click at the upper left corner of the viewport and hold the left mouse button and then drag the cursor to the lower right corner of the viewport. Next, release the left mouse button to set the length and width values; a rectangular spline will be displayed in the viewport, as shown in Figure 5-8.

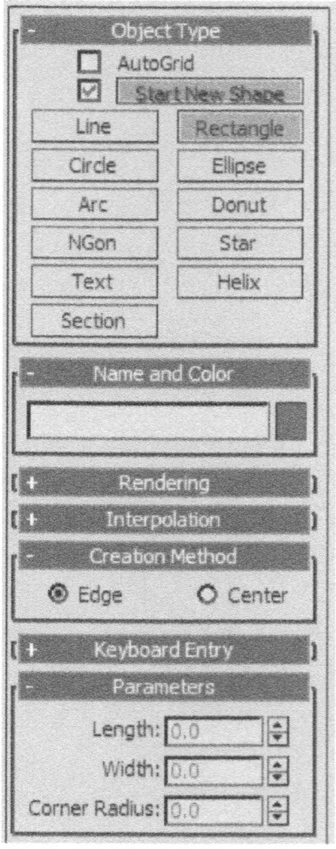

Figure 5-7 Various rollouts to create and modify a rectangular spline

Note

You can also create a square spline, as shown in Figure 5-9, using the same method as described above. The only difference in this case is that you need to hold the CTRL key while dragging the cursor.

Figure 5-8 A rectangular spline

Figure 5-9 A square spline

Creating a Text Spline

Menu bar:	Create > Shapes > Text
Command Panel:	Create > Shapes > Splines > Object Type rollout > Text

The **Text** tool is used to create a spline in the text shape. To create a text spline, activate the viewport in which you want to create it and then choose the **Text** tool from the **Object Type** rollout; various rollouts will be displayed. To create the text dynamically, type the text in the Text text box of the **Parameters** rollout. Next, activate the Front viewport and click on it; the text will be displayed in the viewports, refer to in Figure 5-10.

Figure 5-10 The text in the viewport

Creating a Helix Spline

Menu bar:	Create > Shapes > Helix
Command Panel:	Create > Shapes > Splines > Object Type rollout > Helix

The **Helix** tool is used to create a spline with spirals. To create a helix spline, activate the viewport in which you want to create it and then choose the **Helix** tool from the **Object Type** rollout; the **Name and Color**, **Rendering**, **Creation Method**, **Keyboard Entry**, and **Parameters** rollouts will be displayed. Press and hold the left mouse button and then drag the cursor. Next, release the left mouse button and then drag the cursor up or down to define the height of the helix. Now, click on the viewport and move the cursor up or down to set the second radius of the endpoint. Again, click on the viewport; a helix spline will be created in viewports, as shown in Figure 5-11.

Figure 5-11 A helix spline

TUTORIAL

Tutorial 1

In this tutorial, you will create the model of a book, as shown in Figure 5-12, using splines and extended splines. **(Expected time: 15 min)**

Figure 5-12 The model of a book

The following steps are required to complete this tutorial:

a. Create the project folder.
b. Create a book.
c. Create the spiral binding of the book.
d. Create the text on the book.
e. Save and render the scene.

Creating the Project Folder

Create a new project folder with the name *c05_tut1* at*\Documents\3dsmax2017* and then save the file with the name **c05tut1**, as discussed in Tutorial 1 of Chapter 2.

Creating a Book

In this section, you will create a book using the **Hose** tool from **Extended Primitives**.

1. Activate the Top viewport. Choose **Create > Geometry** in the **Command Panel** and then select **Extended Primitives** from the drop-down list displayed below the **Geometry** button.

Now, choose the **Hose** tool from the **Object Type** rollout.

2. In the Top viewport, press and hold the left mouse button and drag the cursor to define the radius of the hose. Next, release the left mouse button and move the cursor up to define the height of the hose. Now, click on the viewport to get the desired height; a hose is created.

3. Name the hose as *book* in the **Name and Color** rollout and change its color by entering the values as follows:

   Red: **153** Green: **228** Blue: **214**

4. Choose the **Modify** tab in the **Command Panel**. In the **Free Hose Parameters** area of the **Hose Parameters** rollout, set the value **29.798** in the **Height** spinner. Next, in the **Common Hose Parameters** area of the **Hose Parameters** rollout, make sure the **Flex Section Enable** check box is selected. Use default values for other options.

5. In the **Hose Shape** area of the **Hose Parameters** rollout, select the **Rectangular Hose** radio button and then set the values **144.927** and **210.827** in the **Width** and **Depth** spinners, respectively. Use the default values for other options.

6. Choose the **Zoom Extents All** tool to view *book* in viewports, as shown in Figure 5-13.

Creating the Spiral Binding of the Book

In this section, you will create the helix spline for creating the spiral binding of the book.

1. Activate the Front viewport and choose **Create > Shapes** in the **Command Panel**. Make sure that **Splines** is selected in the drop-down list displayed below the **Shapes** button. Now, choose the **Helix** tool from the **Object Type** rollout.

2. Expand the **Keyboard Entry** rollout and enter the values as follows:

   Radius 1: **17.0** Radius 2: **17.0** Height: **205.0**

3. Choose the **Create** button; a helix shape is created in the viewports.

4. Name the helix as *spiral binding* and change its color by entering the values as follows:

   Red: **6** Green: **134** Blue: **6**

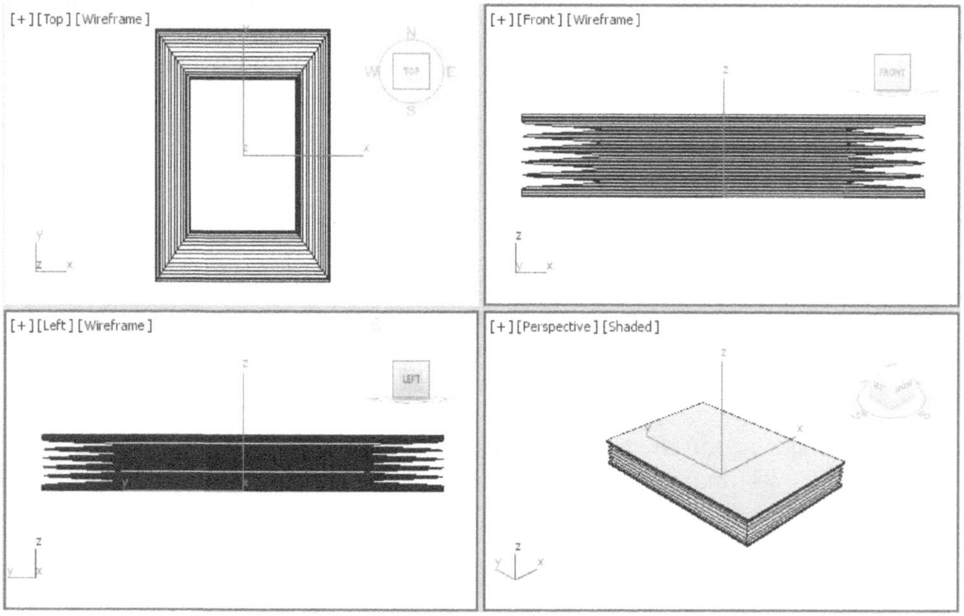

***Figure 5-13** The book geometry after invoking the **Zoom Extents All** tool*

5. Choose the **Modify** tab in the **Command Panel** and then in the **Rendering** rollout, select the **Enable In Renderer** and **Enable In Viewport** check boxes. Make sure the **Radial** radio button is selected and then set the value **1.5** in the **Thickness** spinner. Next, in the **Parameters** rollout, set the value **17.0** in the **Turns** spinner. Choose the **Select and Move** tool and align *spiral binding* with *book* in the viewports, as shown in Figure 5-14.

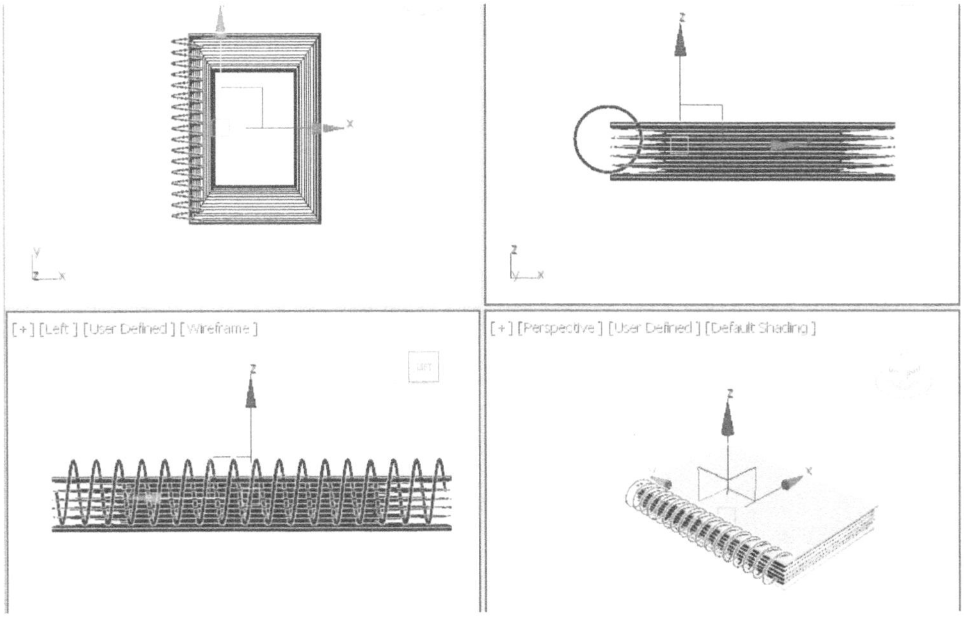

***Figure 5-14** Alignment of spiral binding with book in the viewports*

Creating the Text on the Book

In this section, you will use the **Text** tool to create the text on the book.

1. Choose **Create > Shapes** from the **Command Panel**; the **Splines** option is selected in the drop-down list. Now, choose the **Text** tool from the **Object Type** rollout. Also, select the **AutoGrid** check box in the **Object Type** rollout.

2. In the **Parameters** rollout of the **Text** tool, type **CADCIM** in the **Text** text box and press ENTER. Next, type **3ds Max** and click in the Perspective viewport; the text is displayed in the viewports.

3. Name the text as *text01* and change its color by entering the values as follows:

   Red: **6**   Green: **134**   Blue: **6**

4. Choose the **Modify** tab in the **Command Panel** and then select the **Monotype Corsiva** font type from the drop-down list located at the top of the **Parameters** rollout. Next, choose the center align button and enter the values as follows:

   Size: **23**   Kerning: **1.0**   Leading: **1.0**

5. Choose the **Select and Move** tool and align *text01* in viewports, refer to Figure 5-15.

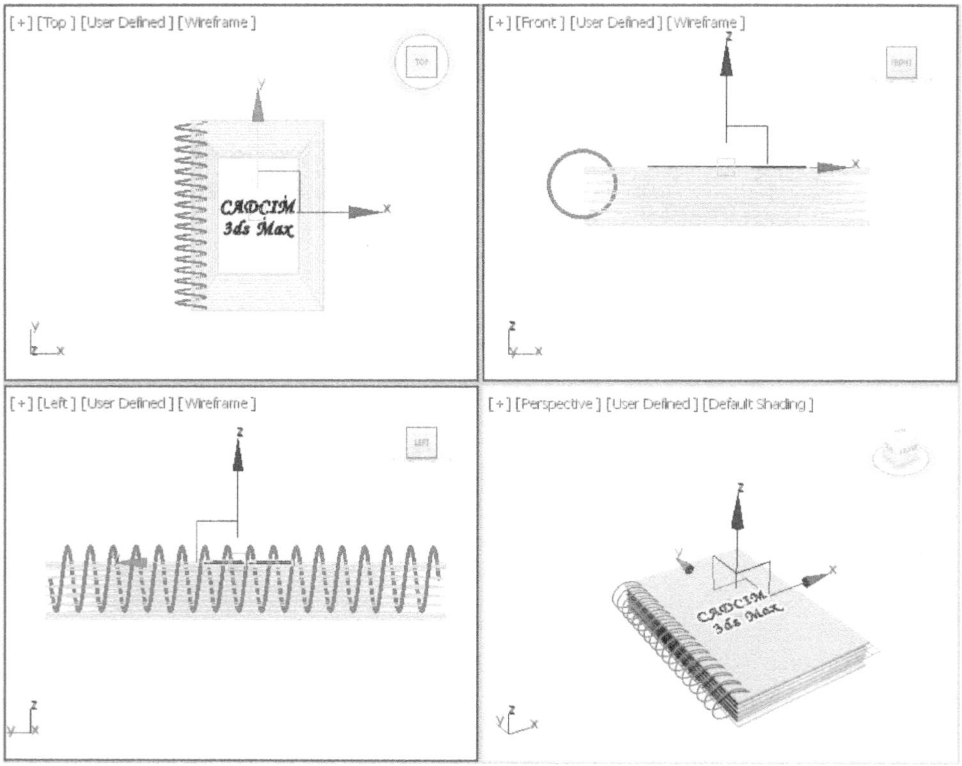

Figure 5-15 *Alignment of text01 on book in viewports*

Next, you need to create a rectangular design over the text.

6. Make sure that the **Splines** option is selected in the drop-down list of the **Command Panel**. Choose the **Rectangle** tool from the **Object Type** rollout. Make sure the **AutoGrid** check box is still selected.

7. Activate the Perspective viewport. Click at the upper left corner of *book,* hold the left mouse button, and then drag the cursor to the lower right corner of *book*. Next, release the left mouse button to set the length and width; a rectangular spline is created in the viewports.

8. Name the rectangle as *rectangular design* and change its color by entering the values as follows:

   Red: **6** Green: **134** Blue: **6**

9. Choose the **Modify** tab in the **Command Panel**. In the **Rendering** rollout, set the value **1.0** in the **Thickness** spinner. Next, set the parameters in the **Parameters** rollout as follows:

   Length: **126.581** Width: **87** Corner Radius: **10.0**

10. Choose the **Select and Move** tool and align *rectangular design* in the viewports, as shown in Figure 5-16.

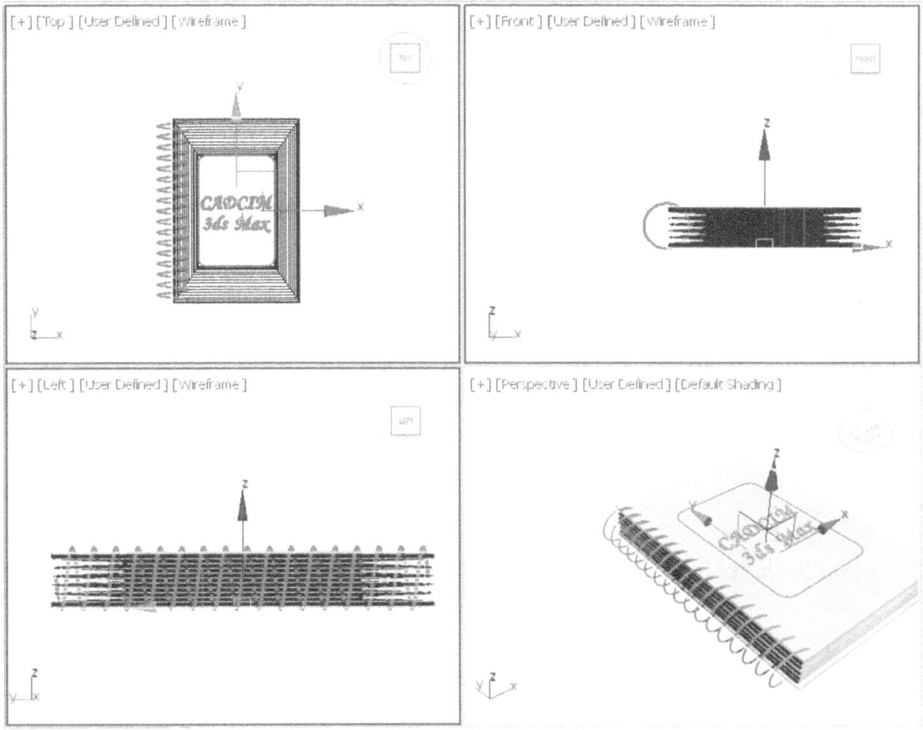

Figure 5-16 Alignment of rectangular design in the viewports

Saving and Rendering the Scene

In this section, you will save the scene and then render it. You can also view the final rendered image of this scene by downloading the *05_3dsmax_2017_rndr.zip* file *www.cadsofttech.com*.The path of the file is as follows: *Textbooks > Animation and Visual Effects > 3ds Max > Autodesk 3ds Max 2017 for Novices*.

1. Change the background color of the scene to white, as discussed in Tutorial 1 of Chapter 2.

2. Choose **Save** from the **Application** menu.

3. Activate the Perspective viewport. Next, choose the **Render Production** tool from the **Main Toolbar**; the **Rendered Frame** window is displayed, showing the final output of *book*, refer to Figure 5-17. Next, close this window.

Figure 5-17 *The final output after rendering*

EXERCISES

The rendered output of the models used in the exercises can be accessed by downloading the *c05_3dsmax_2017_exr.zip* file from *www.cadsofttech.com*.The path of the file is as follows: *Textbooks > Animation and Visual Effects > 3ds Max > Autodesk 3ds Max 2017 for Novices*.

Exercise 1

Create the model of a grid shown in Figure 5-18 using your own dimensions.

(Expected time: 15 min)

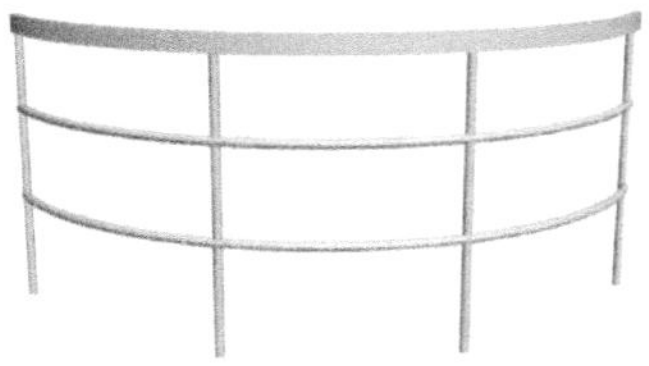

Figure 5-18 *The model of a grid*

Exercise 2

Create the model of a Christmas tree using AEC extended objects and decorate it using various spline objects, as shown in Figure 5-19. **(Expected time: 20 min)**

Figure 5-19 *The Christmas tree model*

Exercise 3

Create the model of an entrance shown in Figure 5-20 using your own dimensions.
(Expected time: 20 min)

Figure 5-20 *The model of an entrance*

This page is intentionally left blank

Chapter 6

Modifying Splines

Learning Objectives

After completing this chapter, you will be able to:

- *Adjust the pivot point of an object*
- *Use the Lathe modifier to modify objects*
- *Modify shapes*
- *Use editable spline objects*

INTRODUCTION

In this chapter, you will learn to modify shapes such as splines and extended splines at advanced level. Also, you will learn about the pivot point and the **Lathe** modifier.

LATHE MODIFIER

The **Lathe** modifier is a shape modifier and is used to create 3D objects by rotating shapes about their axes. The direction of the axis of revolution depends on the pivot point of the shape. To apply the **Lathe** modifier on a shape, create a shape in the Front viewport and select it, as shown in Figure 6-1. Choose the **Modify** tab in the **Command Panel** and then select **OBJECT-SPACE MODIFIERS > Lathe** from the **Modifier List** drop-down list; the 2D shape will be converted into a 3D object and it will be displayed in the viewports, refer to Figure 6-2. When you apply the **Lathe** modifier on a shape, it will be displayed in the modifier stack. The **Parameters** rollout will be displayed below the modifier stack.

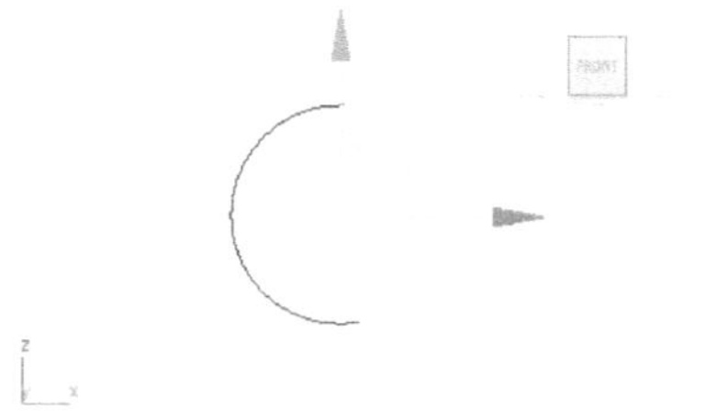

Figure 6-1 *An arc displayed in the Front viewport*

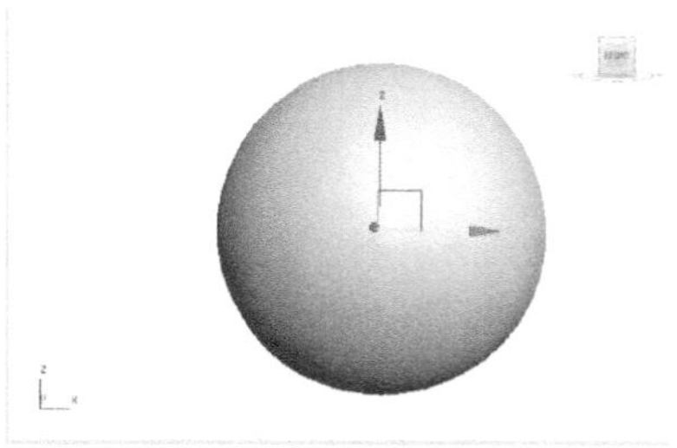

Figure 6-2 *The arc converted into a 3D object after applying the **Lathe** modifier*

MODIFYING THE SHAPES

To modify the shapes in 3ds Max, first create a shape using the tools from **Create > Shapes > Splines/Extended Splines > Object Type** rollout in the **Command Panel** as discussed earlier. Next, select the shape in the viewport and choose the **Modify** tab in the **Command Panel**; the Modify panel will be displayed with various rollouts. Now, you can modify the shapes using one of the following methods:

1. By entering new parameters in the rollouts displayed in the Modify panel. You have already learned about it in the previous chapters.

2. By applying a modifier from the **Modifier List** drop-down list. You will learn about applying modifiers in later Chapters.

3. By converting the shapes into editable splines, which is discussed next.

Converting the Shapes into Editable Splines

You can modify all the spline and extended spline shapes at advanced level by converting them into editable spline objects. The editable spline objects provide the sub-object levels of the shape. These sub-object levels help you to select the sub-objects of the shape in the viewport. You can convert a shape into an editable spline using one of the following methods:

- Create a shape and then select it in the viewport. Next, right-click on the shape; a quad menu will be displayed. Choose the **Convert To** option; a cascading menu will be displayed. Choose **Convert to Editable Spline**; the shape will be converted into editable spline.

- Create a shape and then select it in the viewport. Choose the **Modify** tab in the **Command Panel**. In the modifier stack, right-click on the name of the tool; a shortcut menu will be displayed. Choose the **Editable Spline** option; the shape will be converted into an editable spline.

- Create a shape and then choose the **Modify** tab in the **Command Panel**. Click on the **Modifier List** drop-down list in the **Command Panel** and select **Editable spline** from it; the **Editable Spline** modifier will be applied to the shape in the modifier stack.

Sub-object Levels in Editable Spline

When you convert a spline object into editable spline, the name of the tool in the modifier stack will be replaced by **Editable Spline**. This level is known as the object level. Click on the arrow sign on the left of the **Editable Spline** to view the sub-object levels such as **Vertex**, **Segment**, and **Spline**, as shown in Figure 6-3. On converting a spline object into editable spline, various rollouts such as **Rendering**, **Interpolation**, **Selection**, **Soft Selection**, and **Geometry** will also be displayed in the **Modify** panel, refer to Figure 6-3. These sub-object levels and rollouts are discussed next.

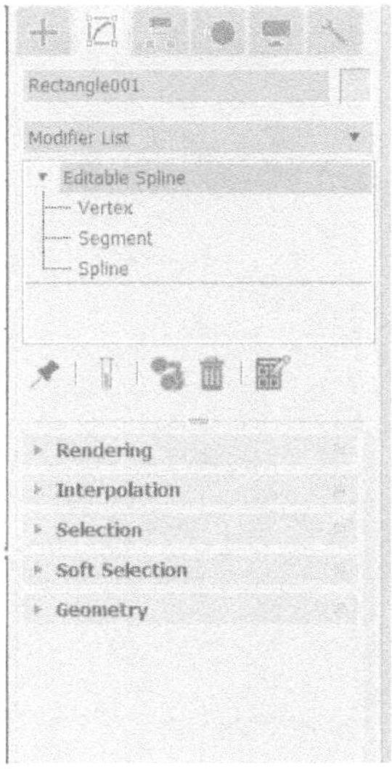

***Figure 6-3** The sub-object levels and rollouts displayed*

Sub-object Levels

The sub-object levels help you to modify the object by moving, rotating, and scaling the sub-objects. To do so, select one of the sub-object levels in the modifier stack; it will be activated and turn yellow. Also, it will prompt you to select the corresponding sub-objects in the viewport. You can select a sub-object in the viewport by clicking on it. If you want to add more sub-objects, then hold the CTRL key and click on the sub-objects that you want to add. You can also drag a selection box around the sub-objects to select them simultaneously. These sub-objects are discussed next.

Vertex

The vertex is a point on the spline. To modify an object using vertices, select the **Vertex** sub-object level; the white and yellow colored vertices of the spline will be displayed in the viewports, refer to Figure 6-4. The yellow colored vertex is the first vertex or the starting point of the spline, refer to Figure 6-4.

There are various types of vertices such as bezier, corner, smooth, and bezier corner vertices, which depend on the type of spline. You can convert a vertex into a bezier, corner, smooth, or bezier corner vertex. To do so, select a vertex in the viewport and right-click on it; a quad menu will be displayed, as shown in Figure 6-5. Choose the type of vertex from the quad menu, refer to Figure 6-10; the vertex will change based on the option chosen from the quad menu.

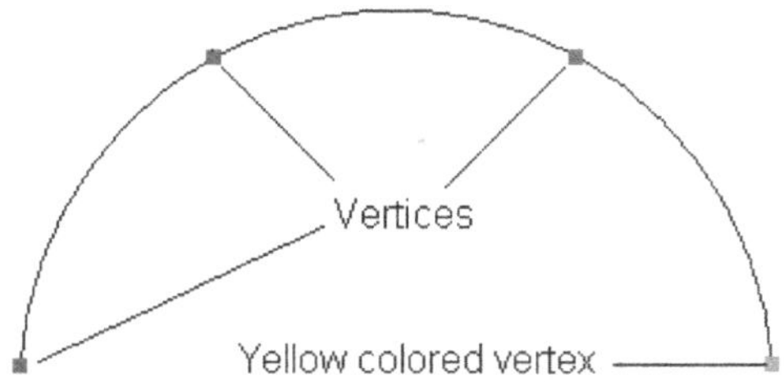

Figure 6-4 *The vertices of an arc spline displayed after selecting the **Vertex** sub-object level*

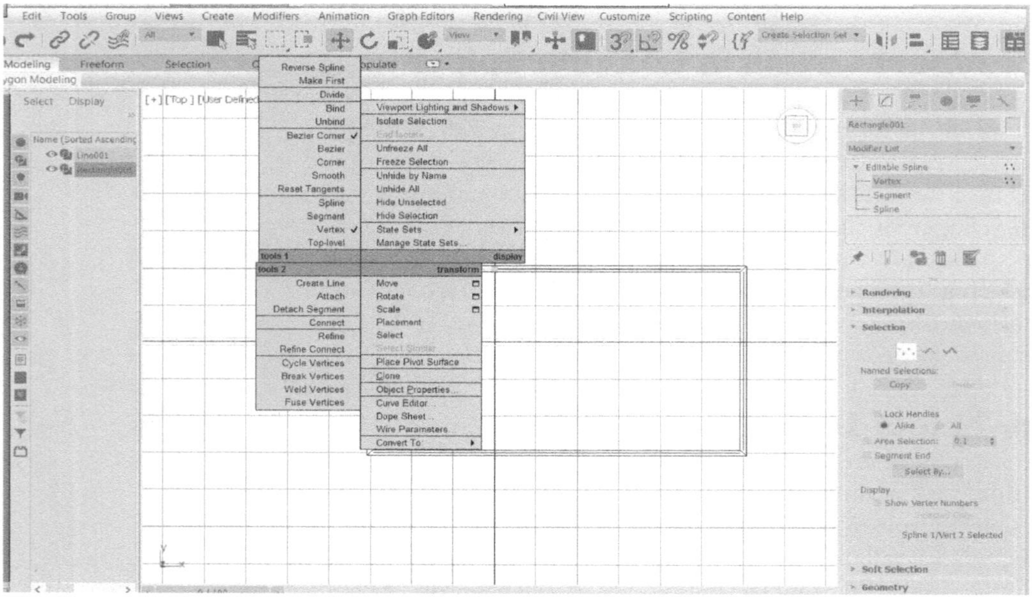

Figure 6-5 *Various types of vertices in a quad menu*

The bezier vertex has bezier handles on both sides of the vertex for adjusting the shape of the spline. The smooth vertex is used to smoothen the curve of the spline. The corner vertex is used to make the corners of the spline. The bezier corner vertex also has bezier handles and is used to adjust the shape of the spline independently on both the sides, refer to Figure 6-6.

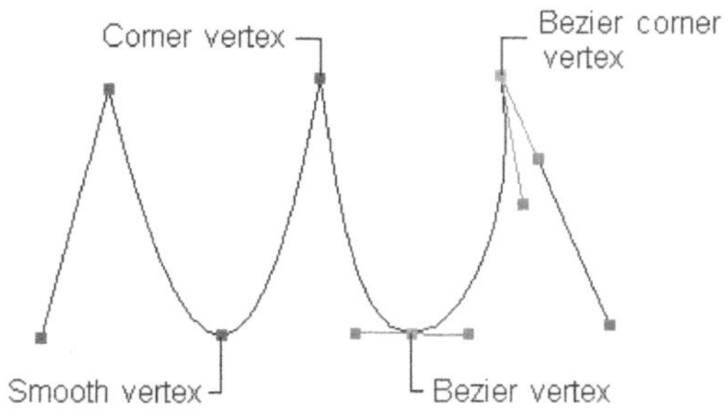

Figure 6-6 *Various types of vertices in a line spline*

Segment

The segment is a line or a curve between two vertices. To modify an object through segments, you need to select the **Segment** sub-object level. On doing so, you will be prompted to select the segments of the object, refer to Figures 6-7 and 6-8.

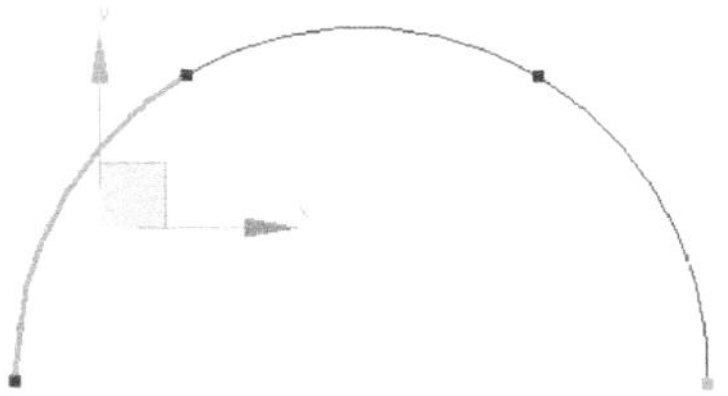

Figure 6-7 *The curved segment selected in an arc spline*

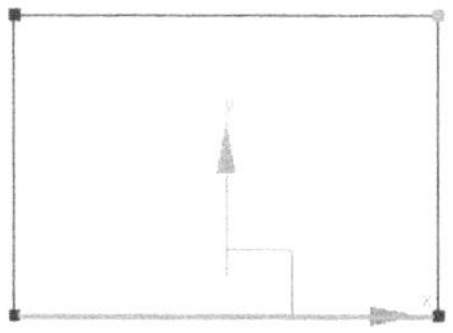

Figure 6-8 *The line segment selected in a rectangular spline*

Spline

A spline consists of all the adjacent segments of an object. To modify an object using the spline, you need to select the **Spline** sub-object level. On doing so, you will be prompted to select all the adjacent segments of the object, as shown in Figure 6-9.

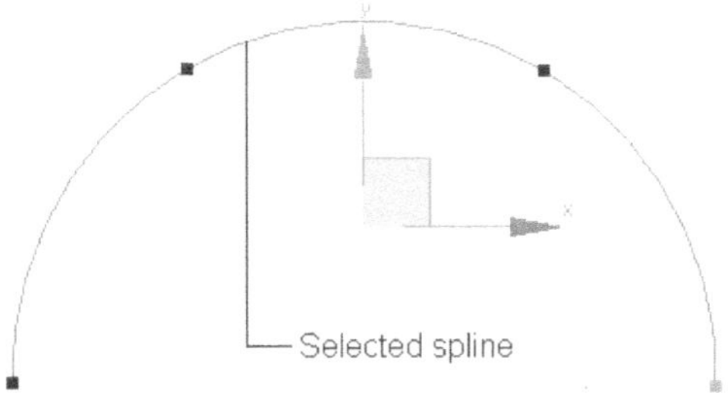

Figure 6-9 *The spline selected in an arc*

Note

*1. By default, the **Line** spline has all the sub-object levels and rollouts. So, you do not need to convert it into editable spline.*

*2. The **Surface Properties** rollout will not be displayed at the **Vertex** sub-object level.*

Rollouts

When you convert a spline object into an editable spline object, various rollouts will be displayed below the modifier stack in the Modify panel. You can modify the spline objects at sub-object levels using these rollouts. These rollouts are discussed next.

Selection Rollout

At the top of the **Selection** rollout, there are three buttons, **Vertex**, **Segment**, and **Spline**, refer to Figure 6-10. These buttons are used to select the sub-objects of a spline and have been discussed in the previous section. Choose one of the buttons to make it active. It is same as selecting the sub-object level in the modifier stack. The check boxes in the **Display** area are used to display the vertex numbers in the viewport. When you select the **Show Vertex Numbers** check box, the vertex numbers will be displayed in the viewport, as shown in Figure 6-11. Select the **Selected Only** check box to display the vertex number only for the selected vertex.

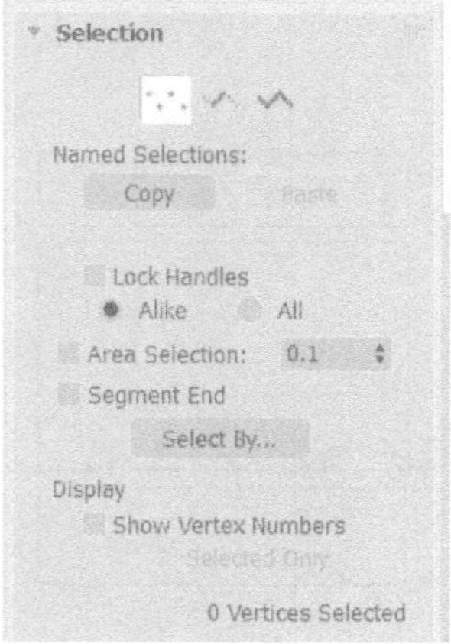

***Figure 6-10** The **Selection** rollout*

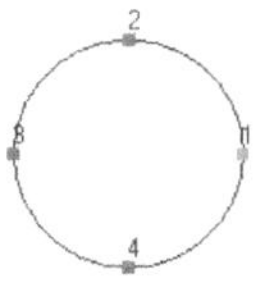

***Figure 6-11** The vertex numbers displayed on selecting the **Show Vertex Numbers** check box*

Soft Selection Rollout

The options in this rollout allow you to soft-select the objects. You can select the **Use Soft Selection** check box to activate the options in this rollout. When you select the sub-objects of an object in the viewport, it shows the falloff in different color combinations. The influence of the colored zone is determined by the value in the **Falloff** spinner. The shape of the falloff is controlled by the value in the **Pinch** and **Bubble** spinners.

Geometry Rollout

The options in this rollout are used to edit the spline at the object and sub-object levels, refer to Figure 6-12. These options are discussed next.

Create Line

The **Create Line** button is used to add a line spline to the selected spline. To do so, select the editable spline object in the viewport and choose the **Modify** tab; all the rollouts will be displayed in the Modify panel. Now, choose the **Create Line** button in the **Geometry** rollout; the button will be activated. Move the cursor in the viewport; its shape will change, as shown in Figure 6-13. Create a new line; it will automatically be attached to the selected spline. Next, right-click in the viewport or choose the **Create Line** button again to exit the command. The **Attach** button is used to attach another spline to the selected spline. When you choose the **Attach Mult.** button, the **Attach Multiple** dialog box will be displayed. It contains the list of all other splines in the scene. Select the shapes in the dialog box that you want to attach and then choose the **Attach** button to attach the shapes.

Insert

The **Insert** button is used to insert additional vertices in the splines to create new segments. To do so, select the editable spline object in the viewport and choose the **Modify** tab; all rollouts will be displayed in the Modify panel. Now, choose the **Insert** button in the **Geometry** rollout; it will get activated. Next, move the cursor over the spline; the shape of the cursor will change, as shown in Figure 6-14. Then, click to create a new vertex and move the cursor to place it. Again, move the cursor and click in the viewport to place the new vertices on the spline and right-click to exit the command.

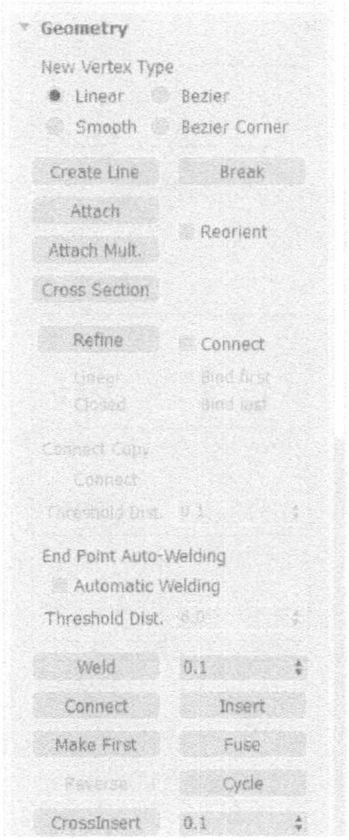

*Figure 6-12 Partial view of the **Geometry** rollout*

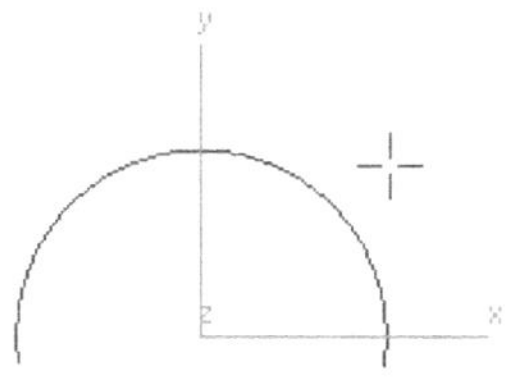

*Figure 6-13 The cursor displayed in the viewport on choosing the **Create Line** button*

*Figure 6-14 The cursor displayed after choosing the **Insert** button*

Break

The **Break** button is available for the **Vertex** and **Segment** sub-object levels. It is used to break a spline at the selected point. If you select the **Vertex** sub-object level, then select one or more vertices in the viewport to break the spline. Next, choose the **Break** button in the **Geometry** rollout; the spline will split at the selected point. Similarly, if you select the **Segment** sub-object level, then choose the **Break** button and click on a point on the selected spline to break it. To view the split splines, select the **Spline** sub-object level in the modifier stack.

Refine and Connect

The **Refine** button and the **Connect** check box are available for the **Vertex** and the **Segment** sub-object levels. The **Refine** button is used to add new vertices to the selected spline. To add new vertices, choose the **Refine** button and move the cursor over the selected spline; the cursor for refining the spline will be displayed. Then, click on a point to add a new vertex. You can repeatedly click on the spline to create more than one vertex. Next, right-click or choose the **Refine** button again to exit the command.

The **Connect** check box on the right of the **Refine** button is used to connect the new vertices by a segment. To do so, select the **Connect** check box, choose the **Refine** button, and then move the cursor over the selected spline; the cursor for refining will be displayed. Now, click on a point to add a new vertex and then click again to create another vertex. Next, right-click to exit the command; a segment will be created to join the two new vertices, as shown in Figure 6-15.

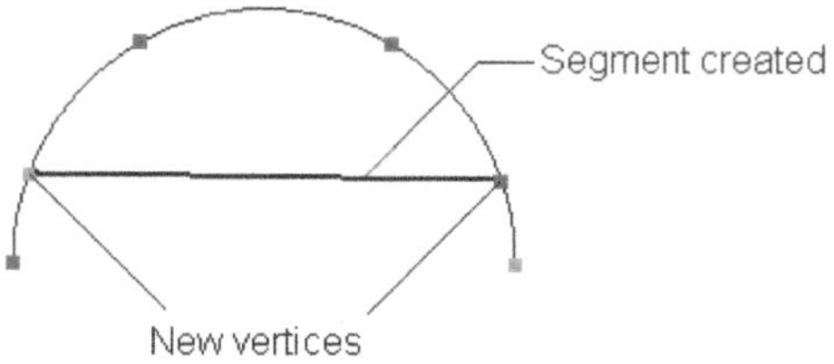

***Figure 6-15** The new vertices and the segment created using the **Refine** button and the **Connect** check box*

Note

*If you insert a vertex using the **Refine** button, then the curve shape of the selected spline will not change. But, if you insert the vertex using the **Insert** button, then the curve shape will change.*

Weld

The **Weld** button is available only for the **Vertex** sub-object level. It is used to join the selected vertices into a single vertex. To do so, select the editable spline object in the viewport and then select the **Vertex** sub-object level; the **Weld** button will be available in the **Geometry** rollout. Next, select the vertices that you want to join together in the viewport and choose the **Weld** button; the vertices will be joined together to create a vertex. The spinner on the right side of the **Weld** button is used to specify the distance between the vertices that can be welded. If the value in the spinner is less than the distance between the vertices, then the vertices will not be welded.

Connect

The **Connect** button is available only for the **Vertex** sub-object level. It is used to connect two end vertices of a spline by creating a segment in between the vertices, refer to Figure 6-16. To do so, select the editable spline object in the viewport. Next, select the **Vertex** sub-object level in the modifier stack, and then choose the **Connect** button in the **Geometry** rollout. Next, move the cursor over one of the end vertices of the spline; the shape of the cursor will change, refer to Figure 6-17. Now, press and hold the left mouse button and drag it to move the cursor over another end vertex; the shape of the cursor will change again and a dotted line will be generated, as shown in Figure 6-18. Now, release the left mouse button; a segment will be created joining the end vertices, as shown in Figure 6-19.

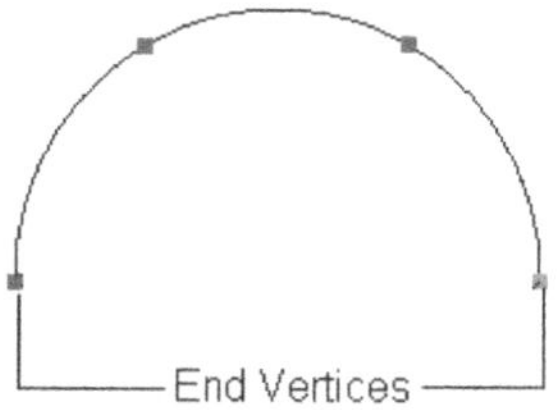

***Figure 6-16** The end vertices in an arc spline*

***Figure 6-17** The cursor over one of the end vertices*

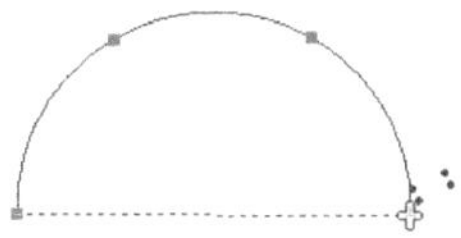

Figure 6-18 The cursor over another end vertex and the dotted line displayed

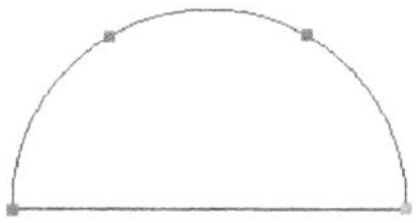

Figure 6-19 The end vertices connected with a segment

Make First

The **Make First** button is used to make any vertex of the spline as its first vertex or the starting point. To do so, select a vertex of the spline in the viewport and choose the **Make First** button in the **Geometry** rollout; the selected vertex will turn yellow and become the first vertex of the spline.

Note

In an open spline such as line, arc, and so on, only the end vertex can be the first vertex.

Fuse

The **Fuse** button is used to move all the selected vertices of a spline to their average center. Select the vertices of a spline in the viewport, as shown in Figure 6-20, and then choose the **Fuse** button in the **Geometry** rollout; the selected vertices will fuse together at their average center, as shown in Figure 6-21.

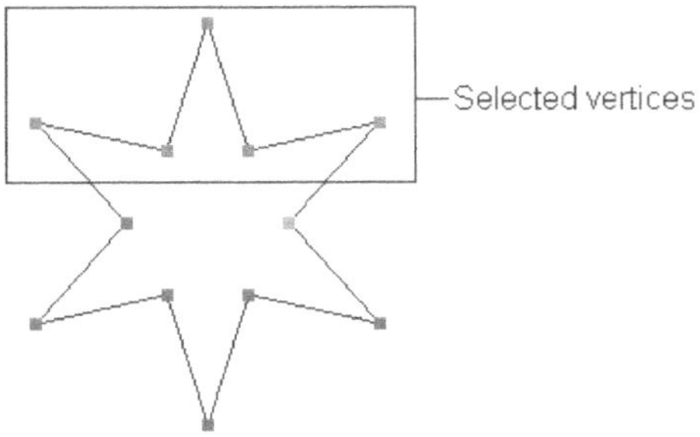

Figure 6-20 The selected vertices inside the rectangular area

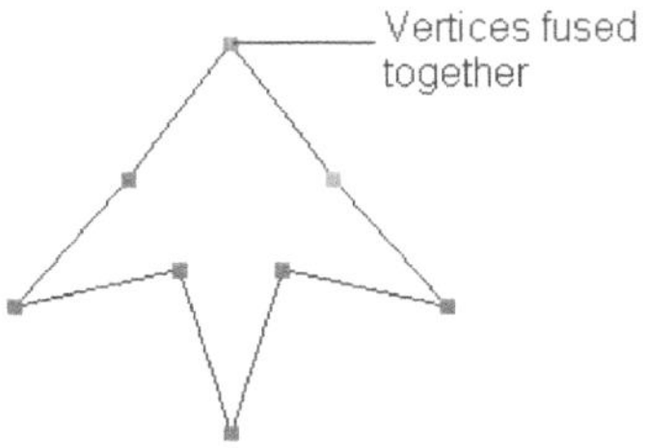

Figure 6-21 The selected vertices fused at their average center

Fillet

The **Fillet** button is used to create round corners in a spline by creating two new vertices. Create a star spline, refer to Figure 6-22. Next, choose the **Fillet** button in the **Geometry** rollout and move the cursor over a vertex; the shape of the cursor will change, as shown in Figure 6-23. Now, press the left mouse button and drag the cursor up; the new vertices with a round corner will be displayed, as shown in Figure 6-24. When you drag the cursor to fillet the vertex, the value of fillet will also be displayed in the spinner on the right side of the **Fillet** button. You can also perform this command by selecting a vertex and entering the value for the fillet in this spinner.

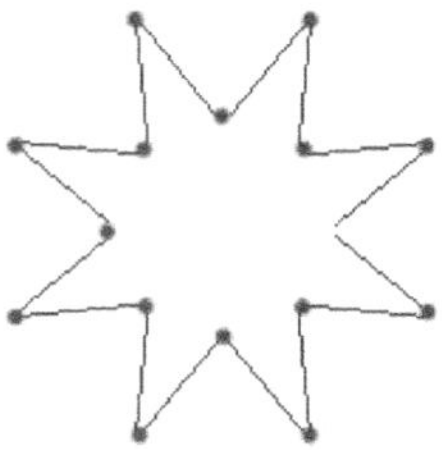

Figure 6-22 The star spline

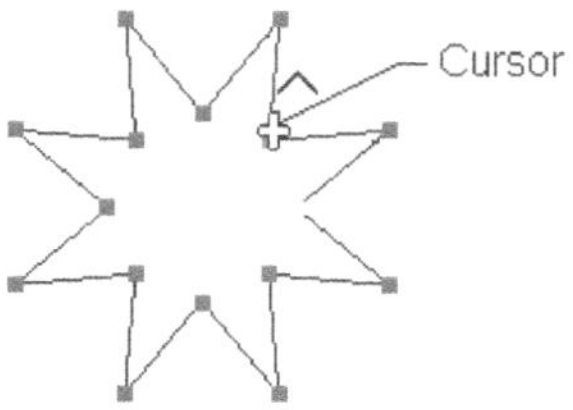

Figure 6-23 The changed shape of cursor displayed after moving the cursor over the selected vertex

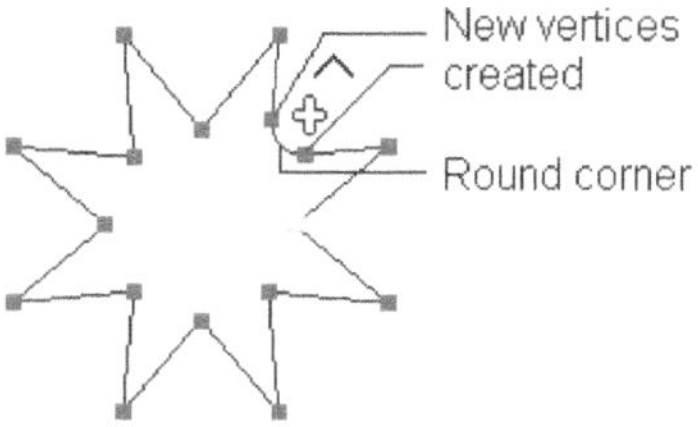

Figure 6-24 The two new vertices and the round corner displayed

Chamfer

The **Chamfer** button is used to create bevel corners in a spline by creating two new vertices. To do so, select the corner vertex of an editable spline object in the viewport. Next, choose the **Chamfer** button in the **Geometry** rollout and move the cursor over the selected vertex; the shape of the cursor will change. Now, press the left mouse button and drag the cursor up; the new vertices with a beveled corner will be displayed, as shown in Figure 6-25. Next, release the left mouse button. You can also chamfer by selecting a vertex and entering the value for the chamfer in this spinner.

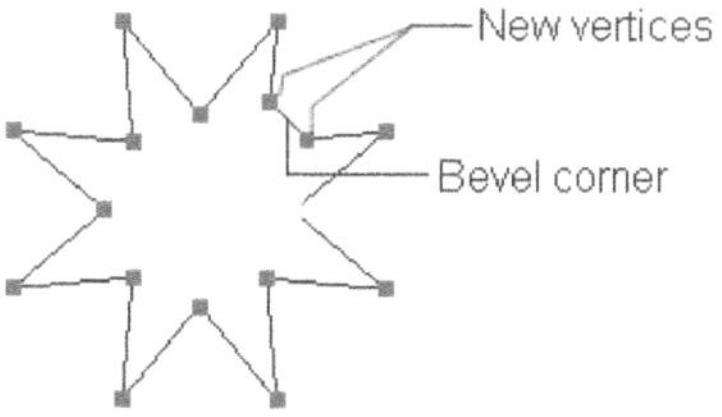

Figure 6-25 The new vertices with a beveled corner

Hide and Unhide All

The **Hide** and **Unhide All** buttons are available for all sub-object levels in the editable spline object. The **Hide** button is used to hide the selected sub-object in the viewport. The **Unhide All** button is used to show all hidden sub-objects in the viewport. To hide the sub-objects of the editable spline object, select them in the viewport and choose the **Hide** button in the **Geometry** rollout. Similarly, to unhide the hidden sub-objects, choose the **Unhide All** button in the **Geometry** rollout.

Delete

The **Delete** button is available at all the sub-object levels and is used to delete the selected sub-objects in the viewport. To do so, select a sub-object in the viewport and choose the **Delete** button; the selected sub-object will be deleted.

Divide

The **Divide** button is available only for the **Segment** sub-object level. It is used to subdivide the selected segment by inserting a number of vertices on it. The number of vertices to be inserted can be set in the spinner on the right side of the **Divide** button. To subdivide a selected segment, select the **Segment** sub-object in the viewport, as shown in Figure 6-26. Next, choose the **Divide** button; the selected segment will be sub-divided, refer to Figure 6-27.

Detach

The **Detach** button is available at the **Segment** and **Spline** sub-object levels. It is used to detach the selected sub-objects from the original one and create a new shape. To do so, select one or more segments of the editable spline object in the viewport, refer to Figure 6-28. Next, choose the **Detach** button in the **Geometry** rollout; the **Detach** dialog box will be displayed, as shown in Figure 6-29. By default, the name of the new shape will be displayed as *Shape001* in the **Detach as** text box. Modify the name if you want and then choose the **OK** button; the detached segment will now behave like a new independent shape, refer to Figure 6-30.

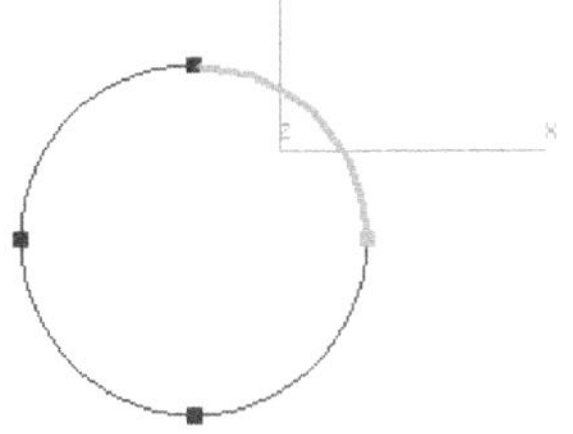

Figure 6-26 A segment selected in the circle spline

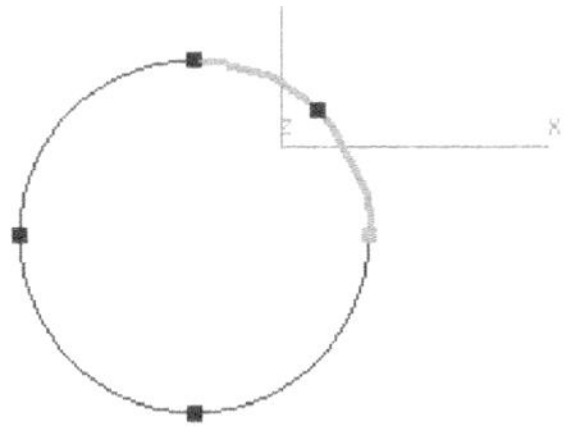

*Figure 6-27 The selected segment sub-divided after using the **Divide** button*

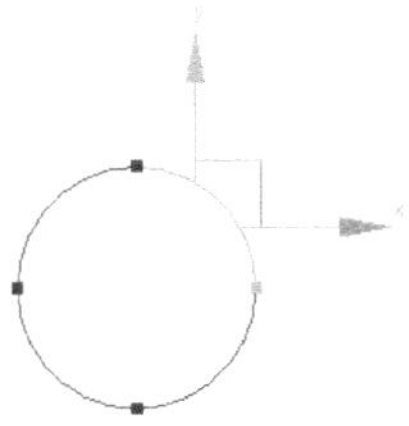

Figure 6-28 A segment sub-object selected in the circle spline

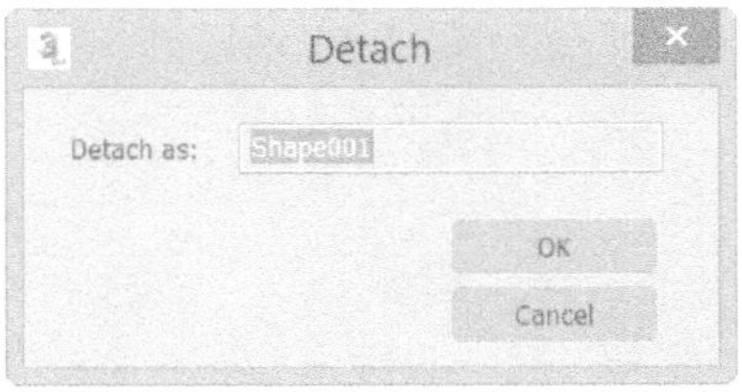

*Figure 6-29 The **Detach** dialog box*

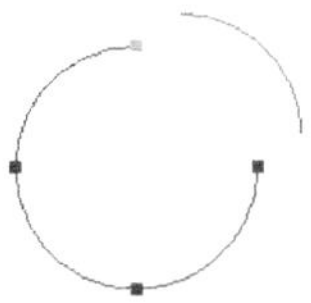

Figure 6-30 The detached segment as a new independent shape

Reverse

The **Reverse** button is available only at the **Spline** sub-object level. It is used to reverse the direction of the spline. To do so, select the editable spline object in the viewport, select the **Spline** sub-object level in the modifier stack, and then select the spline in the viewport, as shown in Figure 6-31. Now, choose the **Reverse** button in the **Geometry** rollout; the direction of the selected spline will get reversed, as shown in Figure 6-32. To view the reversed direction, you need to select the **Show Vertex Numbers** check box in the **Display** area of the **Selection** rollout.

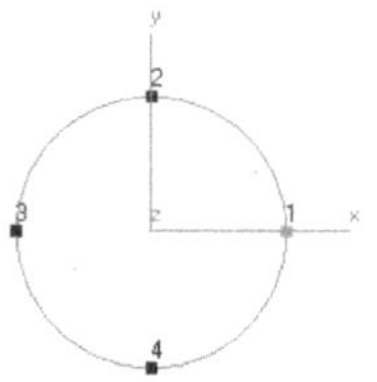

Figure 6-31 *The spline selected in a circle*

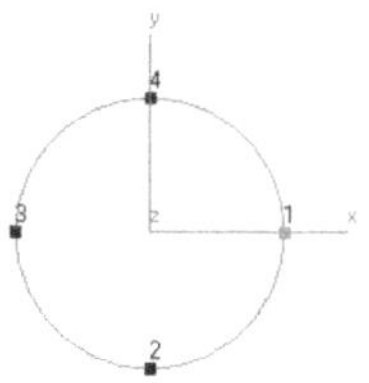

Figure 6-32 *The reversed direction of the selected spline after choosing the* ***Reverse*** *button*

Outline

The **Outline** button is available only at the **Spline** sub-object level. It is used to create a copy of the spline at a distance that you enter in the spinner on the right side of the **Outline** button. To do so, select the spline of the editable spline object in the viewport. Next, choose the **Outline** button; the button will be activated. Next, move the cursor over the selected spline; the shape of the cursor will change, as shown in Figure 6-33. Now, press and hold the left mouse button and move the cursor up or down over the selected spline to create a copy of the spline which behaves like its outline, as shown in Figure 6-34. Now, release the left mouse button. Also, right-click in the viewport or choose the **Outline** button again to exit the command.

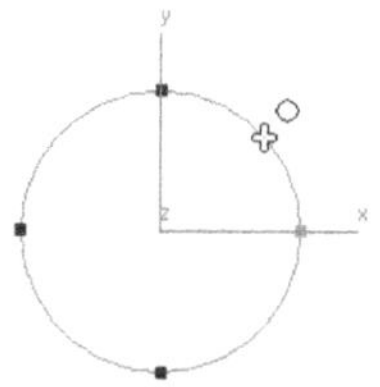

Figure 6-33 *The shape of the cursor changed after choosing the* ***Outline*** *button*

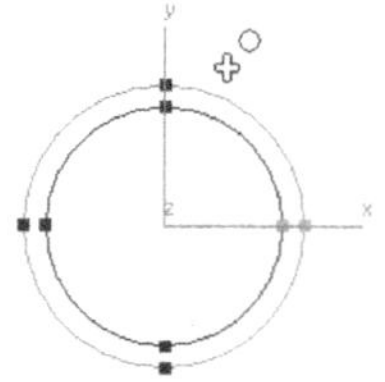

Figure 6-34 *A copy of the spline created as an outline of the selected spline*

Note

If you use the ***Outline*** *button for an open spline such as arc, line, and so on, then the copied spline will be attached to the selected spline. Also, the resultant spline will be a closed one, refer to Figures 6-35 and 6-36.*

Figure 6-35 An open arc spline

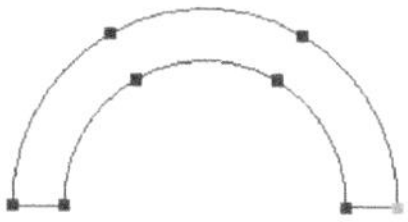

*Figure 6-36 The closed spline after using the **Outline** button on the open spline*

Boolean

The **Boolean** button is available only in the **Spline** sub-object level. You can choose this button to create complex shapes from two spline objects. Note that these two spline objects should be closed, intersect each other, and should be attached. Also, they should be in the same plane. There are three buttons available on the right side of the **Boolean** button, **Union**, **Subtraction**, and **Intersection**.

The **Union** button is used to remove the overlapping portion of two splines and to create a single spline by combining them, refer to Figures 6-37 and 6-38. The **Subtraction** button is used to subtract the overlapping portion of the second spline from the selected spline and remove the extra portion of the second spline, as shown in Figure 6-39. The **Intersection** button is used to remove the non-overlapping portion of the splines and retain the overlapping portion of the splines, as shown in Figure 6-40.

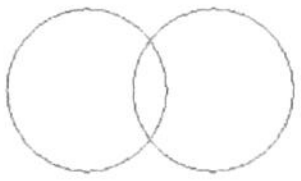

Figure 6-37 The two attached splines overlapping each other

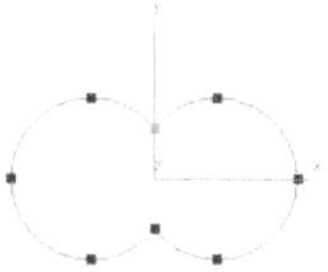

Figure 6-38 The union boolean final output

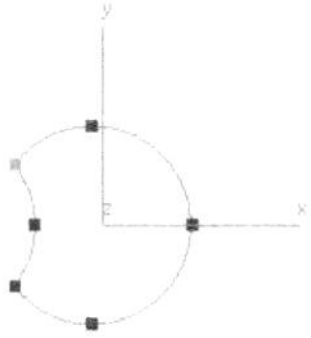

Figure 6-39 The subtraction boolean final output

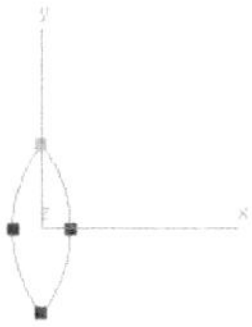

Figure 6-40 The intersection boolean final output

To perform the boolean operation, create two splines and convert them into editable splines and then attach them using the **Attach** button as discussed earlier. Next, select the **Spline** sub-object level in the modifier stack. Now, choose one of the boolean type buttons on the right side of the **Boolean** button in the **Geometry** rollout and then choose the **Boolean** button and move the cursor over the attached splines; the cursor will get changed. Click on the overlapping area; the splines will be modified to form a single spline on the basis of the boolean type button chosen.

Mirror

The **Mirror** button is available only at the **Spline** sub-object level. It is used to move the selected spline in a particular direction. There are three buttons on the right side of the **Mirror** button, **Mirror Horizontally**, **Mirror Vertically**, and **Mirror Both**. These buttons are used to define the direction of movement of the spline. To mirror a spline, select the editable spline in the viewport and then select the **Spline** sub-object level in the modifier stack. Specify the direction of movement of the spline by choosing one of the buttons on the right side of the **Mirror** button. Next, choose the **Mirror** button; the spline will move in the specified direction, refer to Figures 6-41 through 6-43.

Select the **Copy** check box below the **Mirror** button to create a copy of the spline while mirroring, as shown in Figures 6-44 through 6-46. Select the **About Pivot** check box to mirror the spline according to the pivot point of the spline.

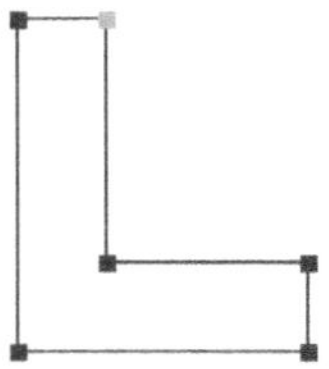

Figure 6-41 *The angle extended spline created to mirror*

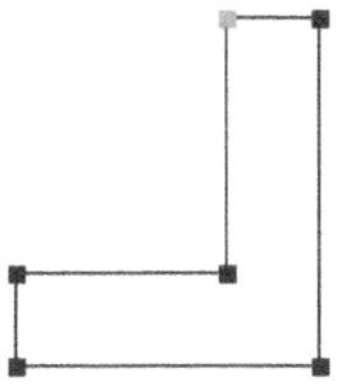

Figure 6-42 *The spline mirrored horizontally*

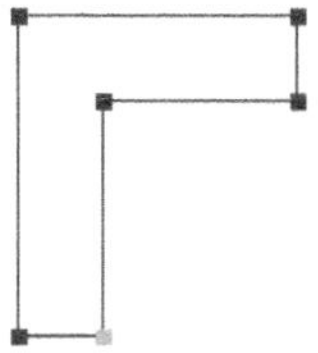

Figure 6-43 *The spline mirrored vertically*

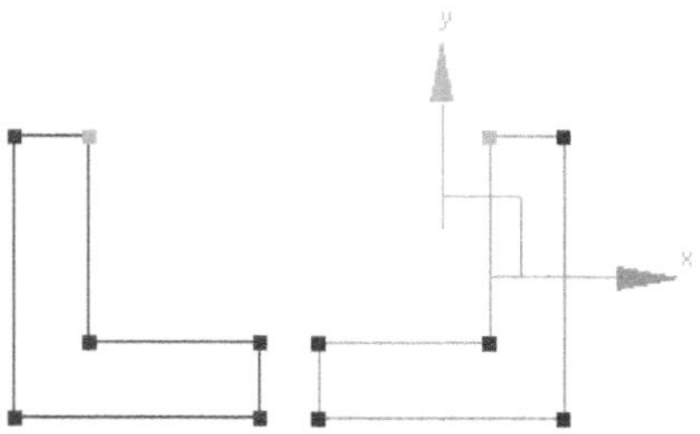

Figure 6-44 *The spline mirrored horizontally after selecting the **Copy** check box*

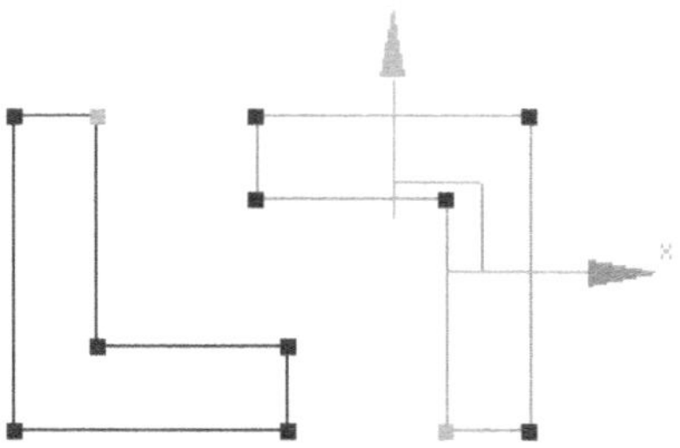

Figure 6-45 *The spline mirrored on both the sides after selecting the **Copy** check box*

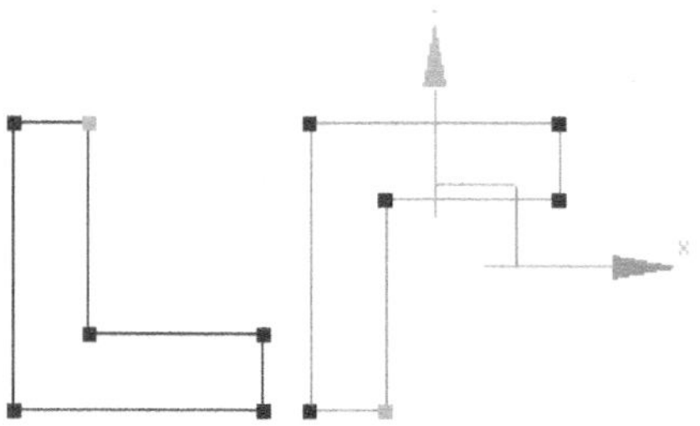

Figure 6-46 *The spline mirrored vertically after selecting the **Copy** check box*

Trim

The **Trim** button is available only at the **Spline** sub-object level. It is used to remove the overlapping segments of the two attached splines. To do so, create two splines, convert them into editable splines, and then attach them using the **Attach** button as described earlier, refer to Figure 6-47. Next, select the **Spline** sub-object level in the modifier stack. Choose the **Trim** button in the **Geometry** rollout and move the cursor over the overlapping segments of the attached splines; the cursor shape will get changed, as shown in Figure 6-52. Now, click on the overlapping segments to remove them, refer to Figure 6-48.

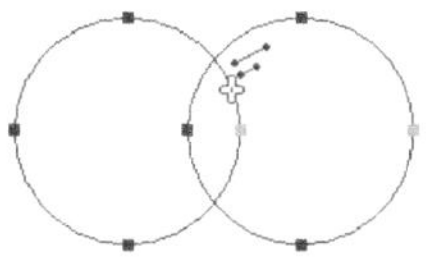

Figure 6-47 *The two attached splines and the cursor over the overlapping segments*

Figure 6-48 *The splines after using the **Trim** button*

Explode

The **Explode** button is available only at the **Spline** sub-object level. It is used to break a spline into a number of splines. The number of broken splines depends on the number of segments of the spline. To break a spline into a number of splines, select the editable spline in the viewport and select the **Spline** sub-object level in the modifier stack. Next, choose the **Explode** button; the spline will be broken into a number of splines, refer to Figures 6-49 and 6-50.

Figure 6-49 *The rectangular spline for explosion*

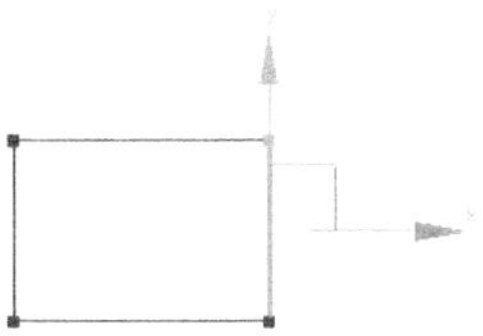

Figure 6-50 *One of the broken splines selected after using the **Explode** button*

There are two radio buttons in the **To** area below the **Explode** button, **Splines** and **Objects**. These buttons are used to select the type of explosion of the spline. By default, the **Splines** radio button is selected, which enables you to break the spline into a number of splines that will be attached together. On selecting the **Objects** radio button, the new splines will be created but they will behave as independent shapes.

Select the **Objects** radio button and then choose the **Explode** button to explode the spline; the **Explode** dialog box will be displayed, as shown in Figure 6-51. Enter a name in the **Object Name** text box and choose the **OK** button; all new shapes will be named according to the name entered in the **Object Name** text box.

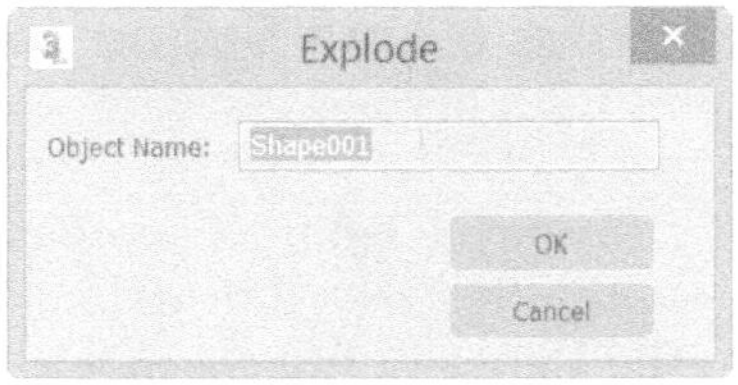

Figure 6-51 *The **Explode** dialog box*

Display Area

The **Display** area will be only active in the **Vertex** and **Segment** sub-object levels. In this area, the **Show selected segs** check box is cleared. As a result, the selected segments will be displayed in the **Segment** sub-object level only. But, if you select the **Show selected segs** check box, the selected segments will be displayed in red color at the **Vertex** sub-object level also.

TUTORIAL

Tutorial 1

In this tutorial, you will create the model of a vase with flower plant, by converting a 2D spline into a 3D mesh object, as shown in Figure 6-52. **(Expected time: 25 min)**

The following steps are required to complete this tutorial:

a. Create the project folder.
b. Create a line spline for the vase.
c. Apply the **Lathe** modifier.
d. Assign the material to the vase.
e. Create a stand for the vase.
f. Save and render the scene.

Figure 6-52 *The model of a vase with flower plant*

Creating the Project Folder

Create a new project folder with the name *c06_tut2* at *\Documents\3dsmax2017* and then save the file with the name *c06tut2*, as discussed in Tutorial 1 of Chapter 2.

Creating a Line Spline for the Vase

In this section, you will create a shape similar to the half portion of the vase using the **Line** tool.

1. Activate the Front viewport and choose **Create > Shapes** in the **Command Panel**; the **Splines** option is selected by default in the drop-down list below the **Shapes** button. Now, choose the **Line** tool from the **Object Type** rollout.

2. In the **Creation Method** rollout, select the **Corner** radio button in both the **Initial Type** and **Drag Type** areas.

3. Create a line spline as discussed in Tutorial 1. This spline should be similar to the half portion of the vase, refer to Figure 6-53.

4. Choose the **Modify** tab and select the **Vertex** sub-object level in the modifier stack; all vertices of the line spline are displayed in the viewport. Next, select the vertices one by one in the viewport and modify the shape by moving them to get a perfect shape.

Applying the Lathe Modifier

In this section, you will apply the **Lathe** modifier to convert the 2D spline into the 3D mesh object.

1. Select the first vertex of the spline in the Front viewport, as shown in Figure 6-54.

2. Choose **Modifiers > Patch/Spline Editing > Lathe** from the menu bar; the 2D line spline is converted into a 3D mesh object.

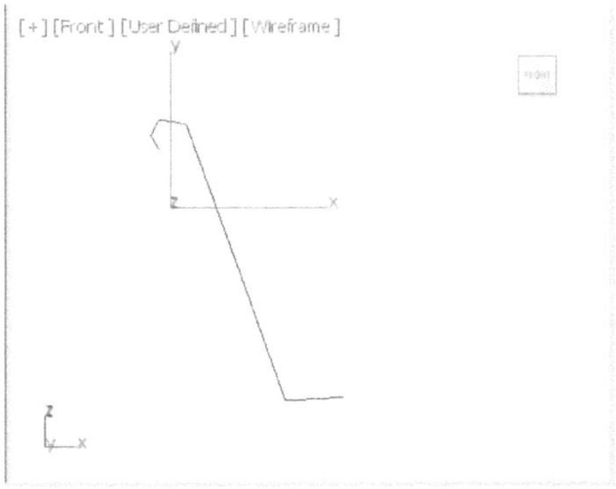

Figure 6-53 *A line spline created for the vase*

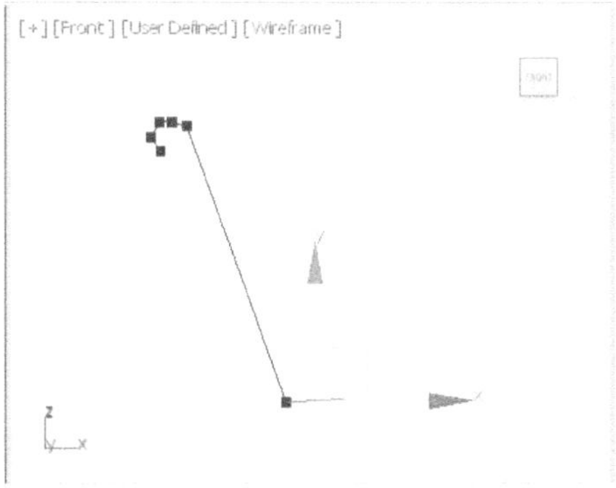

Figure 6-54 *The first vertex of the line spline selected in the Front viewport*

3. Name the new 3D mesh object as *vase*. Also, click anywhere in the viewport to deselect the *vase*.

4. Choose the **Zoom Extents All** tool; the entire *vase* is displayed in the viewports, as shown in Figure 6-55.

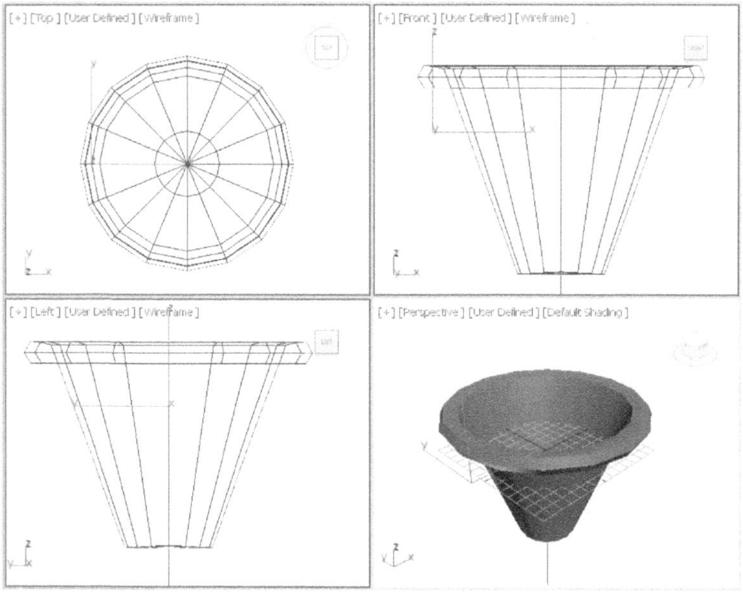

***Figure 6-55** The vase geometry after applying the **Lathe** modifier*

Assigning Material to Vase

In this section, you will assign the material to *vase* using the **Material Editor** tool as discussed in Tutorial 1 of this chapter.

1. Choose **Rendering > Material Editor > Compact Material Editor** from the menu bar; the **Material Editor** dialog box is displayed.

2. Select the **2-Sided** check box in the **Shader Basic Parameters** rollout of the **Material Editor** dialog box. This enables you to view both sides of *vase* in the final output. Also, in the **Blinn Basic Parameters** rollout, choose the **Diffuse** color swatch; the **Color Selector: Diffuse Color** dialog box is displayed. In this dialog box, enter the values as follows:

   Red: **177** Green: **88** Blue: **27**

   Choose the **OK** button in the **Color Selector: Diffuse Color** dialog box; the selected color is displayed in the **Diffuse** color swatch.

3. Make sure that *vase* is selected in the viewports. Next, choose the **Assign Material to Selection** button in the **Material Editor** dialog box; the selected material is assigned to *vase* in the viewport. Next, choose the **Show Shaded Material in Viewport** button in the same dialog box to view the material in viewport. Next, close the **Material Editor** dialog box.

   Next, you need to use the **Foliage** tool to put a plant in *vase*.

4. Activate the Top viewport. Choose **Create > Geometry** in the **Command Panel**; a drop-down list is displayed below the **Geometry** button. Select **AEC Extended** from the drop-down list and then choose the **Foliage** tool from the **Object Type** rollout.

5. In the **Favorite Plants** rollout, double-click on the **Society Garlic** foliage; the foliage is displayed in the viewports.

6. In the **Name and Color** rollout, name the foliage to *plant* and press ENTER.

7. Make sure that *plant* is still selected. Choose the **Modify** tab in the **Command Panel**; the **Parameters** rollout is displayed.

8. In the **Parameters** rollout, set a value in the **Height** spinner to adjust the height of the *plant* according to *vase*.

9. Choose the **Select and Move** tool and align *plant* with *vase* in the viewports, refer to Figure 6-56.

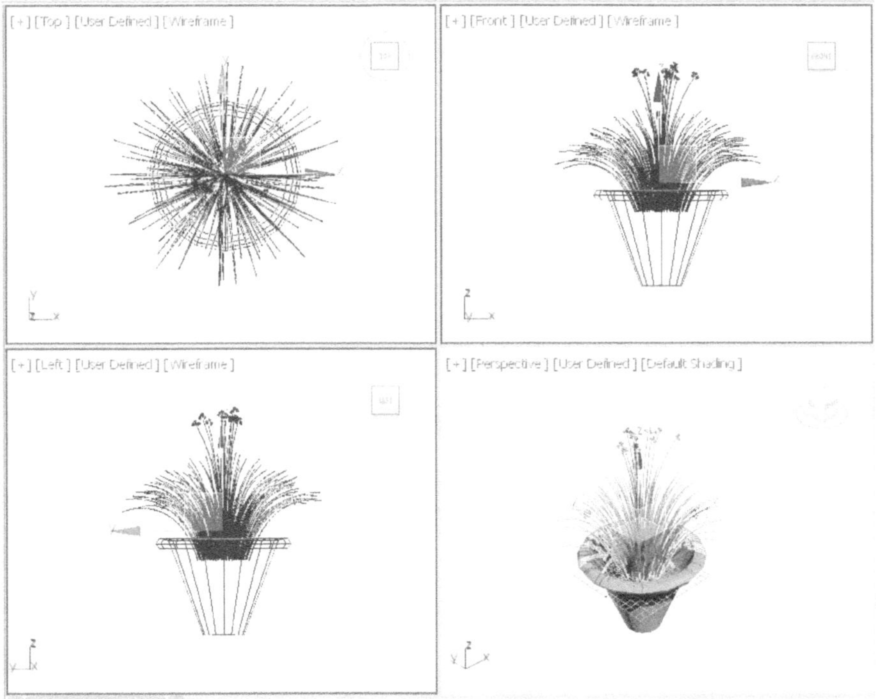

Figure 6-56 *Alignment of plant with vase in viewports*

10. Select *plant* and *vase* in the viewports and group them as *flower vase*.

    Next, you need to create a stand for *flower vase*.

Creating a Stand for the Flower Vase

In this section, you will create a stand for *flower vase*.

1. Select *flower vase* and right-click on it; a quad menu is displayed. Choose the **Hide Selection** option from it; *flower vase* disappears temporarily from the viewports. Now, you can work easily in the viewports.

2. Activate the Top viewport. Choose **Create > Shapes** in the **Command Panel**; the **Splines** option is selected by default in the drop-down list below the **Shapes** button. Now, choose the **Circle** tool from the **Object Type** rollout and create a circular spline.

3. In the **Parameters** rollout, set the radius of the circle to **11.662** in the **Radius** spinner.

   You may need to modify the radius according to the scene in the later section.

4. Expand the **Rendering** rollout and select the **Enable In Renderer** and **Enable In Viewport** check boxes. Select the **Radial** radio button, if it is not already selected. Also, make sure that the **Auto Smooth** check box is selected. Now, enter the values as follows:

   Thickness: **2.0** Sides: **15** Angle: **0.99**

5. The circular spline is automatically named as *Circle001*. Now, change its color by entering the values as follows:

   Red: **193** Green: **193** Blue: **193**

6. Create another circle in the Top viewport. It is automatically named as *Circle002*.

7. In the **Parameters** rollout, set the radius value of *Circle002*. This radius should be double the radius of *Circle001*.

8. Change the color of *Circle002* to black.

9. Choose the **Zoom Extents All** tool and align both the circles using the **Select and Move** tool, as shown in Figure 6-57.

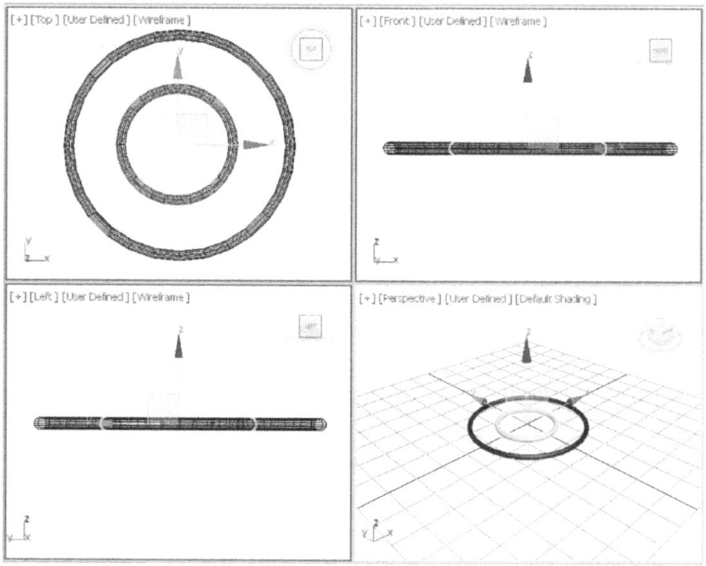

Figure 6-57 *Alignment of the circles in viewports*

10. Choose the **Line** tool from **Create > Shapes > Splines > Object Type** rollout.

11. In the **Creation Method** rollout, select the **Corner** radio button in the **Initial Type** and **Drag Type** areas.

12. Activate the Top viewport. Next, create four line splines to join both the circles, as shown in Figure 6-58.

13. Assign black color to all line splines.

14. Select the line splines and both the circles in any viewport by dragging a selection box around them, and then group them as *upper portion*.

15. Select *upper portion* and hide it temporarily as described earlier.

    Next, you need to create the legs for the stand using the **Line** tool.

16. Activate the Front viewport. Next, choose the **Line** tool from **Create > Shapes > Splines > Object Type** rollout.

17. In the **Creation Method** rollout, select the **Smooth** radio button in both the **Initial Type** and **Drag Type** areas.

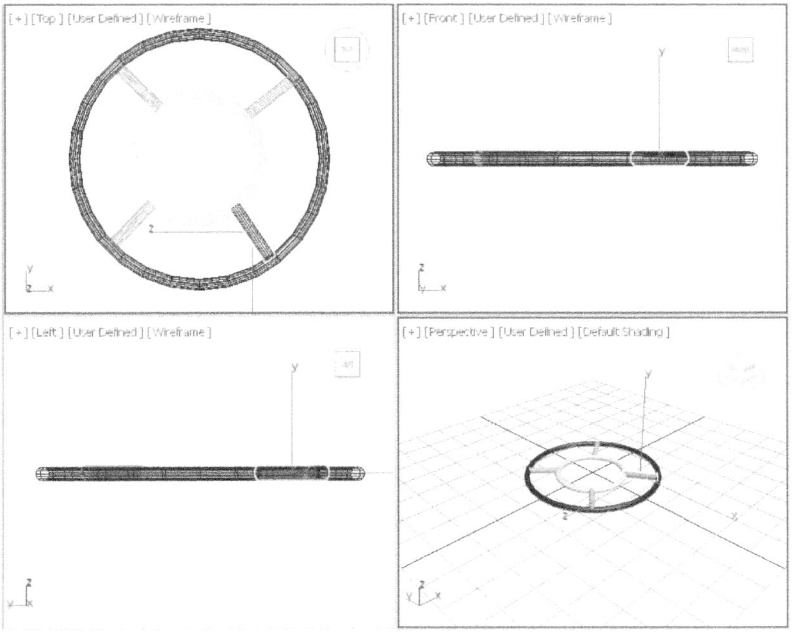

Figure 6-58 *The line splines created to join both the circles*

18. Create a line spline in the Front viewport, as shown in Figure 6-59. It uses the same parameters from the **Rendering** rollout that you had set earlier.

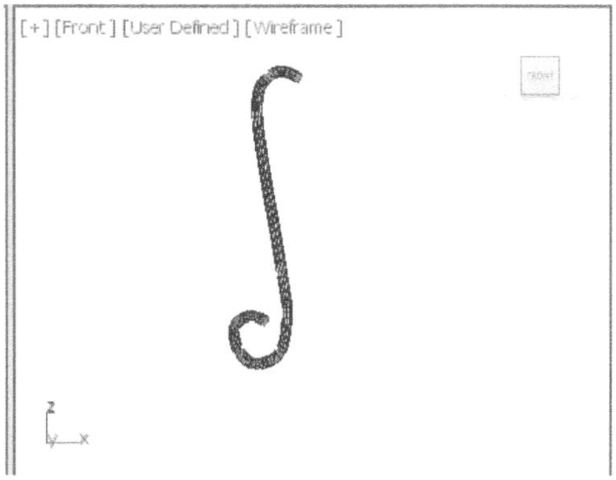

Figure 6-59 *The line spline for the leg of stand*

19. In the **Rendering** rollout, make sure that the **Enable In Renderer** and **Enable In Viewport** check boxes are selected. Select the **Rectangular** radio button and enter the values as follows:

    Length: **2.0** Width: **1.5** Angle: **0.0**

20. Name the line spline as *leg01* and change its color to black.

21. Right-click in any of the viewports; a quad menu is displayed. Choose the **Unhide by Name** option; the **Unhide Objects** dialog box is displayed. Select *upper portion* from the list given in this dialog box and choose the **Unhide** button; *upper portion* is now displayed in the viewport. Choose the **Zoom Extents All** tool.

Note
You can resize leg01 according to the size of the upper portion using the ***Select and Uniform Scale*** *tool.*

22. Align *upper portion* and *leg01* in viewports, as shown in Figure 6-60.

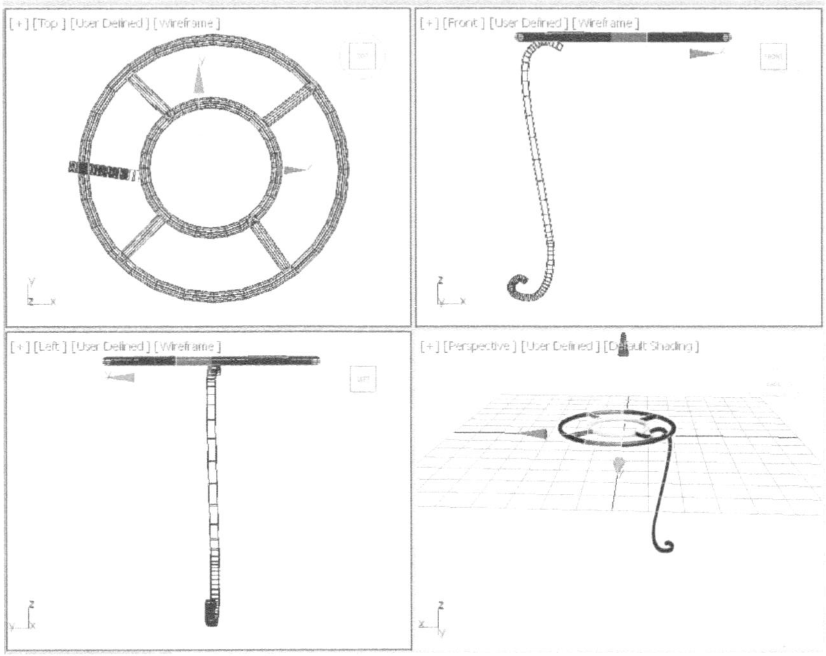

Figure 6-60 *Alignment of upper portion and leg01 in viewports*

Next, you need to copy the *leg01* to create other legs.

23. Create three copies of *leg01*; they are automatically named as *leg002*, *leg003*, and *leg004*.

24. Align *leg01*, *leg002*, *leg003*, and *leg004* using the **Select and Move** and **Select and Rotate** tools, as shown in Figure 6-61.

25. Select all the objects in the viewport and group them as *vase stand*.

26. Right-click on the viewport; a quad menu is displayed. Choose the **Unhide All** option; all the hidden objects are displayed in the viewports.

27. Align *flower vase* and *vase stand* and choose the **Zoom Extents All** tool, refer to Figure 6-62.

Note

You can adjust the size of the flower vase according to the size of the vase stand using the ***Select and Uniform Scale*** *tool.*

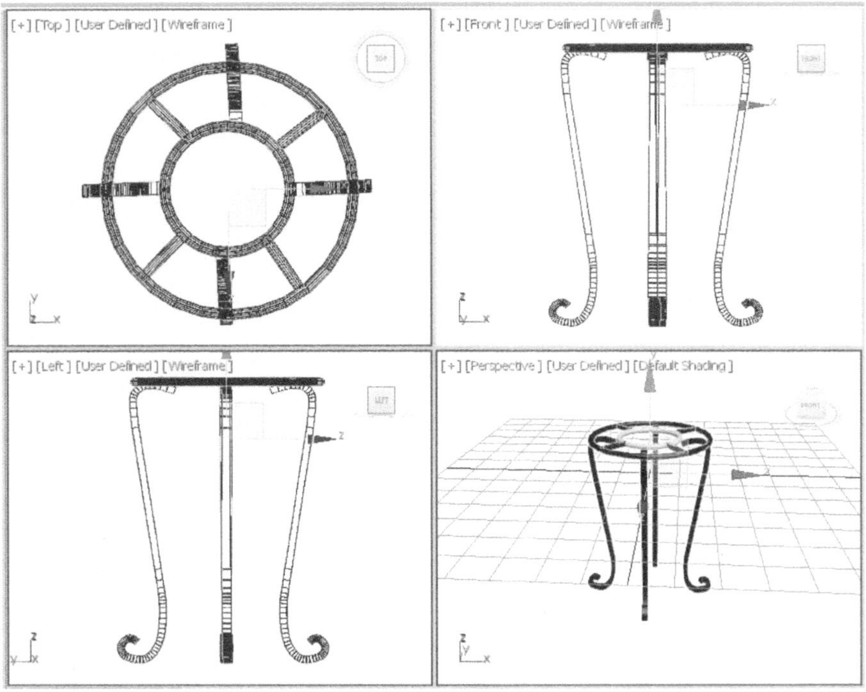

Figure 6-61 *Alignment of leg01, leg002, leg003, and leg004 in viewports*

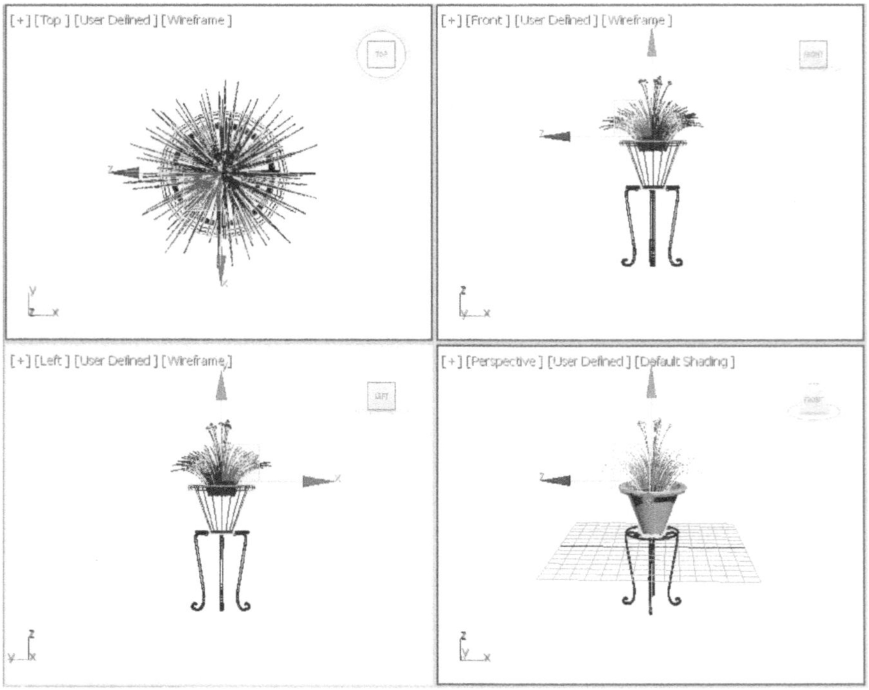

Figure 6-62 *Alignment of flower vase and vase stand in viewports*

Saving and Rendering the Scene

In this section, you will save the scene and then render it. You can also view the final rendered image of this model by downloading the *c06_3dsmax_2017_rndr.zip* file from *www.cadsofttech.com*. The path of the file is as follows: *Textbooks > Animation and Visual Effects > 3ds Max > Autodesk 3ds Max 2017 for Novices*.

1. Change the background color of the scene to white, as discussed in Tutorial 1 of Chapter 2.

2. Choose **Save** from the **Application** menu.

3. Activate the Perspective viewport. Next, choose the **Render Production** tool from the **Main Toolbar**; the **Rendered Frame** window is displayed. This window shows the final output of the scene, refer to Figure 6-62.

EXERCISES

The rendered output of the models used in the Exercise 1 to Exercise 3 can be accessed by downloading the *c06_3dsmax_2017_exr.zip* from *www.cadsofttech.com*.The path of the file is as follows: *Textbooks > Animation and Visual Effects > 3ds Max > Autodesk 3ds Max 2017 for Novices*.

Exercise 1

Create a pair of cocktail glasses, as shown in Figure 6-63, using your own dimensions.

(Expected time: 15 min)

Figure 6-63 *The pair of cocktail glasses*

Exercise 2

Create a set of vases, as shown in Figure 6-64, using your own dimensions.

(Expected time: 15 min)

Figure 6-64 *Set of vases*

Exercise 3

Create a candle stand, as shown in Figure 6-65, using your own dimensions.

(Expected time: 15 min)

Figure 6-65 *A candle stand*

Saving and Rendering the Scene

In this section, you will save the scene and then render it. You can also view the final rendered image of this model by downloading the *c06_3dsmax_2017_rndr.zip* file from *www.cadsofttech.com*. The path of the file is as follows: *Textbooks > Animation and Visual Effects > 3ds Max > Autodesk 3ds Max 2017 for Novices*.

1. Change the background color of the scene to white, as discussed in Tutorial 1 of Chapter 2.

2. Choose **Save** from the **Application** menu.

3. Activate the Perspective viewport. Next, choose the **Render Production** tool from the **Main Toolbar**; the **Rendered Frame** window is displayed. This window shows the final output of the scene, refer to Figure 6-62.

EXERCISES

The rendered output of the models used in the Exercise 1 to Exercise 3 can be accessed by downloading the *c06_3dsmax_2017_exr.zip* from *www.cadsofttech.com*.The path of the file is as follows: *Textbooks > Animation and Visual Effects > 3ds Max > Autodesk 3ds Max 2017 for Novices*.

Exercise 1

Create a pair of cocktail glasses, as shown in Figure 6-63, using your own dimensions.

(Expected time: 15 min)

Figure 6-63 *The pair of cocktail glasses*

Exercise 2

Create a set of vases, as shown in Figure 6-64, using your own dimensions.

(Expected time: 15 min)

Figure 6-64 *Set of vases*

Exercise 3

Create a candle stand, as shown in Figure 6-65, using your own dimensions.

(Expected time: 15 min)

Figure 6-65 *A candle stand*

Chapter 7

Materials and Maps

Learning Objectives

After completing this chapter, you will be able to:

- *Use the Material Editor dialog box*
- *Understand the types of materials and maps*
- *Create materials and maps*
- *Assign materials to objects*

INTRODUCTION

In this chapter, you will learn to assign various materials and maps to the objects in a scene to make it more realistic. Materials are used to describe the reflection or transmission of light of an object. Maps are used to simulate texture, and apply designs, reflection, refraction, and other effects. To assign materials and maps, you need to use the **Material Editor** tool from the **Main Toolbar**, which is discussed next.

MATERIAL EDITOR

Menu bar:	Rendering > Material Editor > Compact/Slate Material Editor
Main Toolbar:	Material Editor flyout > Compact/Slate Material Editor
Keyboard:	M

In 3ds Max, there are two types of material editors, Slate Material Editor and Compact Material Editor. You can use any of these material editors to create new materials and maps, modify the existing materials, apply these new materials and maps to the models in the scene, and create a new environment for the scene. The Slate Material Editor uses nodes and wiring to graphically display the design of the materials created by you. The Compact Material Editor is easy and stripped down version of Slate Material Editor.

To create a new material, press the M key; the **Slate Material Editor** dialog box will be displayed, as shown in Figure 7-1. Next, if you choose **Modes > Compact Material Editor** from the material editor menu bar, the **Material Editor** dialog box will be displayed, as shown in Figure 7-2. The title bar of the **Material Editor** dialog box displays the name and number of the current material, which gets changed depending on the material that you select. The **Material Editor** dialog box is divided into various areas, refer to Figure 7-2. These areas are discussed next.

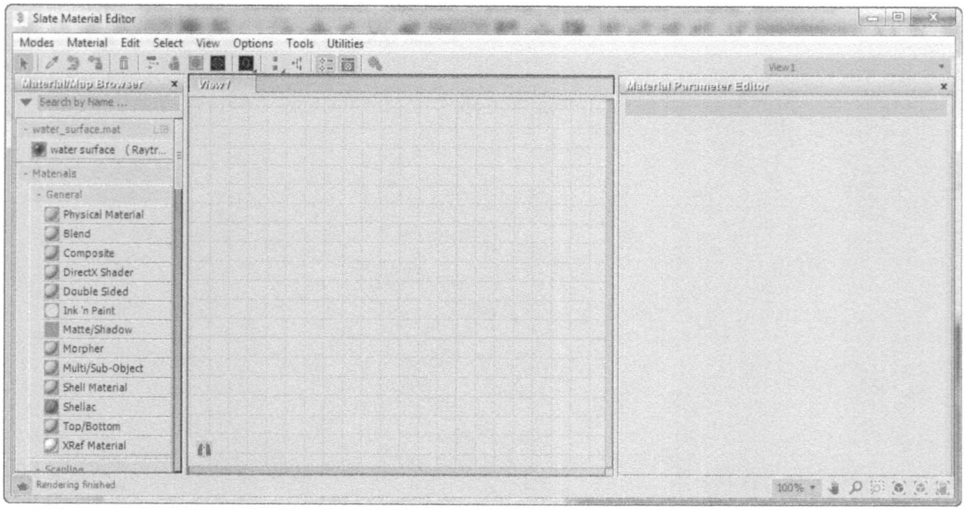

*Figure 7-1 The **Slate Material Editor** dialog box*

Note

*When you press the M key, the material editor that you opened last in 3ds Max will be displayed. However, if you press the M key for the first time in 3ds Max, then the **Slate Material Editor** will be displayed.*

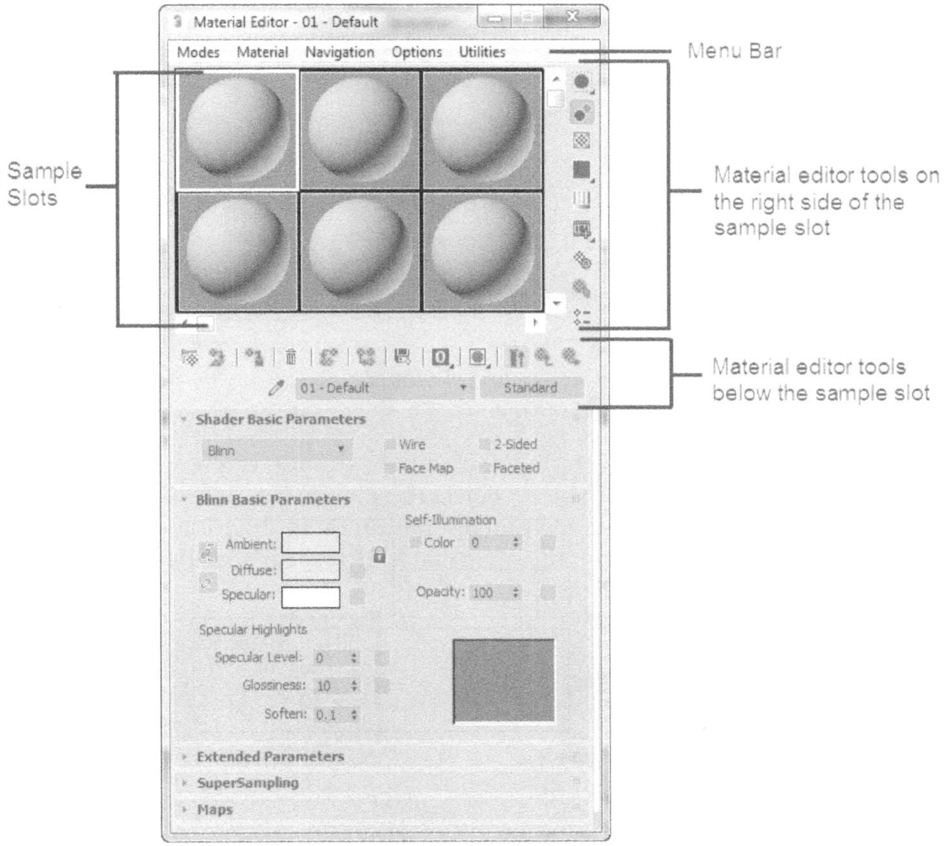

*Figure 7-2 The **Material Editor** dialog box*

Menu Bar

The menu bar is located at the top of the **Material Editor** dialog box. It consists of various pull-down menus such as **Material**, **Navigation**, and so on. The options in these pull-down menus are used to choose various tools. These tools can also be invoked using the options available below as well as on the right side of the sample slots in the **Material Editor** dialog box, refer to Figure 7-2.

Sample Slots

The sample slots in the **Material Editor** dialog box are used to preview the material or the maps that are assigned to the objects in the viewport. By default, there are 6 slots which can be extended to 24 slots and each sample slot represents a single material or map. To view other sample slots, move the scroll bar available on the right, and just below the sample slots. You can also increase the number of visible sample slots. To do so, select one of the sample slots; it will be surrounded by a white colored boundary and will get activated. Next, right-click on the active sample slot; a shortcut menu will be displayed, as shown in Figure 7-3. By default, the **3x2 Sample Windows** option is chosen in the shortcut menu. Next, choose the required option; the number of visible sample slots will be increased accordingly, refer to Figures 7-4 and 7-5.

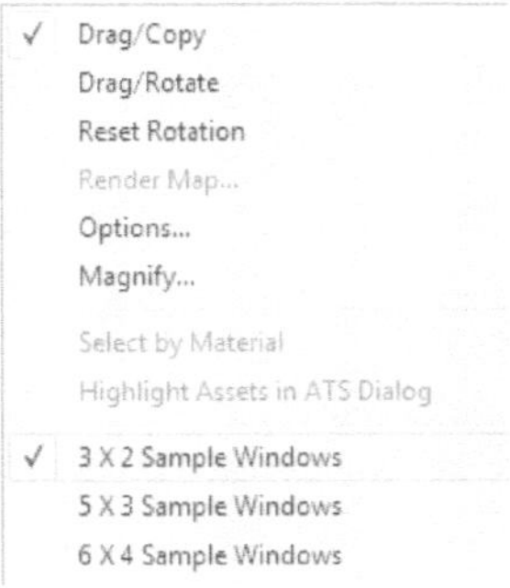

***Figure 7-3** The shortcut menu displayed*

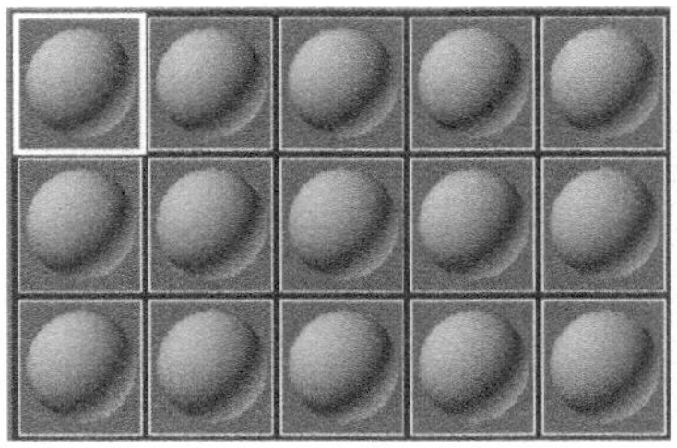

***Figure 7-4** The sample slots after choosing the **5x3 Sample Windows** option*

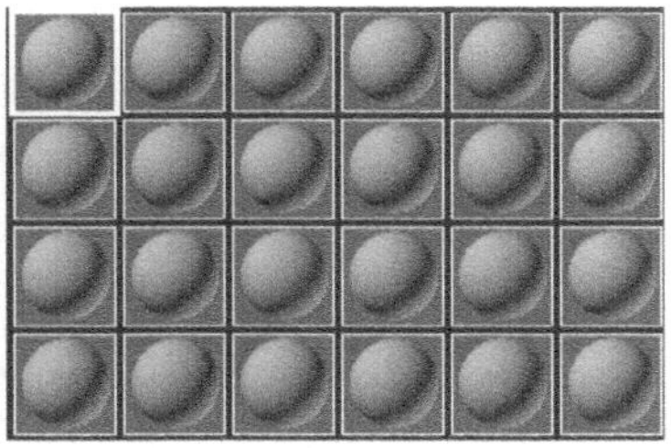

***Figure 7-5** The sample slots after choosing the **6x4 Sample Windows** option*

In the shortcut menu, the **Drag/Copy** option is chosen by default. It is used to copy the material or the map from one sample slot to another. Select the sample slot, whose material or map you want to copy. Next, press and hold the left mouse button on it, drag it to another sample slot, and then release the left mouse button; the same material or the map will be copied to the other sample slot, as shown in Figures 7-6 and 7-7.

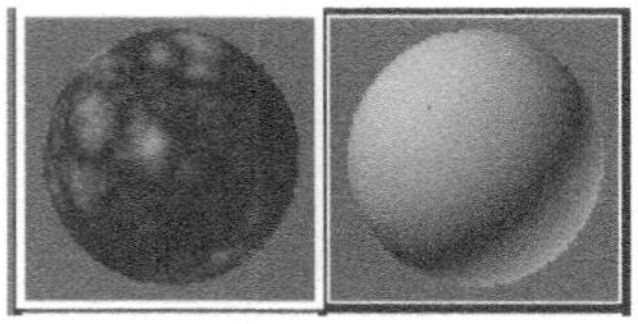

***Figure 7-6** Two sample slots in the **Material Editor** dialog box*

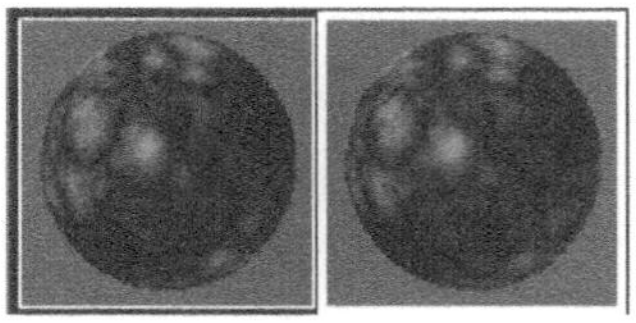

***Figure 7-7** The map/material of one sample slot copied to the other*

On choosing the **Drag/Rotate** option in the shortcut menu, the shape of the cursor will change. Now, you can view the material map on the sample slot by rotating it. The **Reset Rotation** option in the shortcut menu is used to reset the sample slot to its default orientation. By default, the **Render Map** option in the shortcut menu is not activated. It will be activated on assigning a map to the sample slot. This option is used to render the map in the selected sample slot by creating a bitmap or an *avi* file (if the map is animated). When you choose the **Options** option, the **Material Editor Options** dialog box will be displayed. The options in this dialog box are used to modify the sample slots.

Choose the **Magnify** option; the selected sample slot will be displayed in an enlarged window, as shown in Figure 7-8. Alternatively, double-click on the selected sample slot to display it in an enlarged window. To resize the window, move the cursor at the bottom of the window; the cursor will change to a double-headed arrow. Now, drag the cursor to resize the window.

***Figure 7-8** The sample slot displayed in a separate window*

Material Editor Tools

The material editor tools are used to modify the default settings of the selected sample slot, and apply the materials or the maps to the objects. These tools are available below as well as on the right side of

the sample slots in the **Material Editor** dialog box. The most commonly used tools are:
Get Material
Assign Material to Selection
Reset Map/Mtl to Default Settings
Put to Library
Show Shaded Material in Viewport
Go to Parent
Background
Sample UV Tiling
Pick Material From Object

MATERIALS

In this section, you will learn about the **Material Editor** tool that helps you to create materials and assign them to the objects to make them more realistic. The materials assigned to the objects give the best results with lights, about which you will learn in the later chapters. In 3ds Max, you can create and assign various materials to the objects using the **Material Editor** tool from the **Main Toolbar**.

Choose the **Material Editor** tool from the **Main Toolbar**; the **Material Editor** dialog box will be displayed. Now, choose the **Material Type** button on the right side of the **Material Name** drop-down list, refer to Figure 7-9; the **Material/Map Browser** dialog box will be displayed. Next, expand the **Materials > General** rollout, if it is not already expanded. In this list, there are some materials that can combine other materials in them. These are known as Compound materials. The Compound materials in this list are **Blend**, **Composite**, **Double-Sided**, **Morpher**, **Multi/Sub-Object**, **Shellac**, and **Top/Bottom**. Select the material as per your requirement from the list and then choose the **OK** button; the name of the selected material will be displayed on the **Material Type** button in the **Material Editor** dialog box. Also, the rollouts related to that material will be displayed in the **Material Editor** dialog box. The material can be modified by using the parameters available in these rollouts. The most commonly used materials are discussed next.

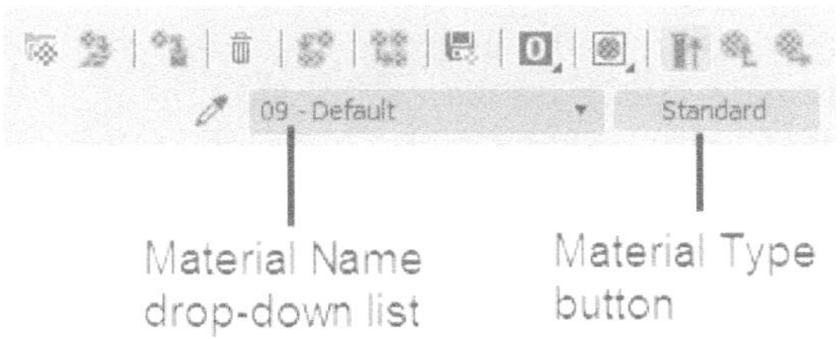

Figure 7-9 *The **Material Name** drop-down list and the **Material Type** button*

Note
*You can also get the default materials using the **Get Material** tool in the **Material Editor** dialog box, as discussed earlier in this chapter.*

Standard Material

By default, the **Standard** material is selected in the **Material Editor** dialog box. Various rollouts are also displayed below the **Material Type** button to create new materials or modify the properties of the current material. The most commonly used rollouts are discussed next.

Architectural Material

The **Architectural** material is used to provide a highly realistic view of an object when rendered with photometric lights and radiosity, about which you will learn in the later chapters. To apply the **Architectural** material, choose the **Material Type** button on the right side of the **Material Name** drop-down list; the **Material/Map Browser** dialog box will be displayed. Choose the **Architectural** material from the **Material/Map Browser** dialog box; the **Architectural** label will be displayed on the **Material Type** button. Also, various rollouts such as **Physical Qualities**, **Special Effects**, and so on will be displayed below the **Architectural** button. You can use these rollouts to create new materials and modify the physical properties of those materials.

Raytrace Material

The **Raytrace** material creates highly refractive and reflective surfaces. When this material is assigned to an object in the scene, the materials/maps of other objects in the scene get reflected on it, as shown in Figure 7-10. It is an advance material and is used to provide realistic raytraced images. It also creates raytraced reflections and refractions, and takes a longer time to render. On selecting the **Raytrace** material, various rollouts will also be displayed below the **Material Type** button to create new materials and modify the properties of those materials. The **Raytrace Basic Parameters** rollout is the most commonly used rollout.

Composite Material

The **Composite** material is similar to the **Blend** material and is used to mix ten different materials. You can assign this mixed material as a single material. On choosing the **Material Type** button, the **Material/Map Browser** dialog box will be displayed. Select the **Composite** material and choose the **OK** button; the **Replace Material** message box will be displayed. Select the **Discard old material?** radio button to remove the material of the selected sample slot. Else, select the **Keep old material as sub-material?** radio button to retain the material of the selected sample slot as a sub-material in the **Composite** material. Choose the **OK** button; the **Composite** material will be displayed in the **Material Type** button with a number of rollouts.

Double Sided Material

The **Double Sided** material is used to assign two different materials to the front and back faces of an object, refer to Figures 7-11and 7-12. Choose the **Double Sided** material from the **Material/Map Browser** dialog box; the **Replace Material** message box will be displayed. Select the **Discard old material?** radio button from this dialog box to remove the material of the selected sample slot. Else, select the **Keep old material as sub-material?** radio button to retain the material of the selected sample slot as the sub-material in the **Double Sided** material. Choose the **OK** button;the **Double Sided** material will be displayed in the **Material Type** button with a number of rollouts.

Figure 7-10 *The object with the **Standard** material*

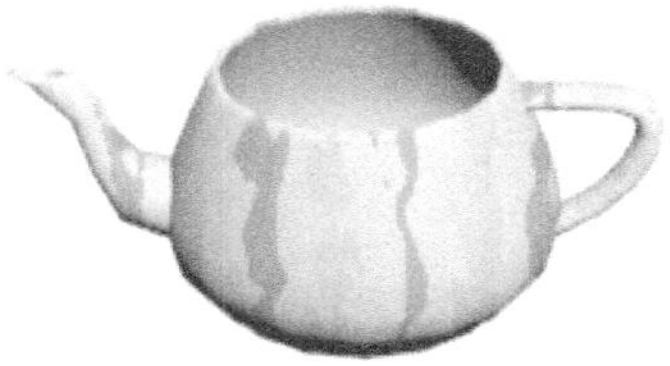

Figure 7-11 *The object with the **Double Sided** material*

Multi/Sub-Object Material

The **Multi/Sub-Object** material is used to assign different materials to an object at the sub-object level, as shown in Figure 7-12. Select the object to which the **Multi/Sub-Object** material is to be assigned. Convert the object to editable poly and then select the **Polygon** sub-object level. Next, select the group of polygons in the viewport, refer to Figure 7-13. Now, expand the **Polygon Material IDs** rollout and enter the value **1** in the **Set ID** spinner and press ENTER. Then, invert the selection and enter the value **2** in the **Set ID** spinner. Next, press ENTER; refer to Figure 7-14.

Figure 7-12 *The object with the **Multi/Sub-Object** material*

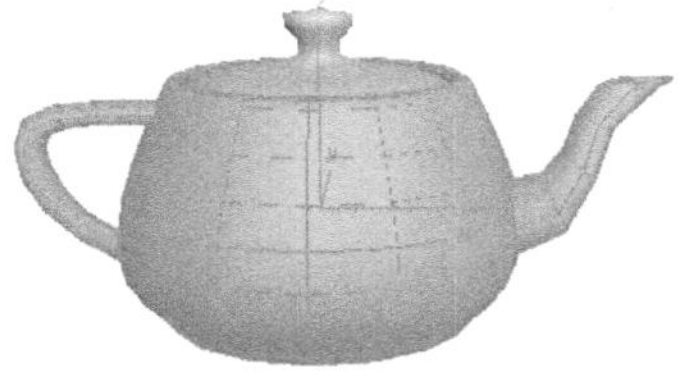

Figure 7-13 *The object with group of polygons selected*

Figure 7-14 *The object with remaining polygons selected*

Note that multiple numbers of IDs can be assigned to an object depending on the number of materials to be assigned to it. Once IDs are set to different parts of the object, choose the **Multi/ Sub-Object** material from the **Material/Map Browser** dialog box; the **Replace Material** message box will be displayed. Select the **Discard old material?** radio button from it and choose the **OK** button; the **Multi/Sub-object** material will be displayed in the **Material Type** button. Also, the **Multi/Sub-Object Basic Parameters** rollout will be displayed, refer to Figure 7-15.

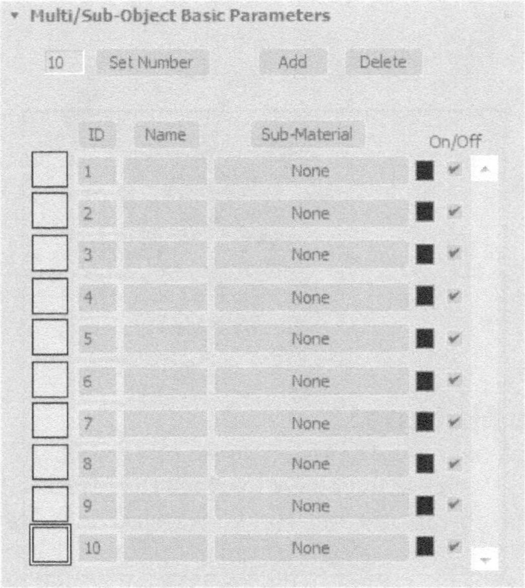

***Figure 7-15** The **Multi/Sub-Object** Basic Parameters rollout*

Top/Bottom Material

The **Top/Bottom** material is used to assign two different materials on the top and bottom portions of the object. To do so, choose the **Top/Bottom** material from the **Material/Map Browser** dialog box; the **Replace Material** message box will be displayed. Select the **Discard old material?** radio button from this message box to remove the material of the selected sample slot. Else, select the **Keep old material as sub-material?** radio button to retain the material of the selected sample slot as the sub-material in the **Top/Bottom** material. Choose the **OK** button; the **Top/Bottom** material will be displayed in the **Material Type** button with a number of rollouts.

Physical Material

The **Physical Material** is a modern layered based material that is suitable for physically-based workflow. This material is compatible with both the mental ray and ART renderers. It also preserves the energy and ensures that the sum of various shading components does not exceed 100%. The light energy is only scattered or absorbed but not created. This material comes with many presets that you can access from the **Presets** rollout of the **Material Editor** dialog box. Presets provide you the starting point to create various materials.

MAPS

In 3ds Max, maps are the default images assigned to a material. They are used to create texture for the objects to make them appear more realistic. In this section, you will learn about assigning the maps to the objects. When maps are assigned to the objects, the objects demonstrate best effects on applying lights.

TUTORIAL

Tutorial 1

In this tutorial, you will create and assign material to the objects in the scene, refer to Figure 7-16. **(Expected time: 30 min)**

Figure 7-16 *The textured scene*

The following steps are required to complete this tutorial:

a. Create the project folder.
b. Open the file.
c. Create material for bench.
d. Create material for street lamp.
e. Create material for dustbin.
f. Create material for pots and plants.
g. Save and render the scene.

Creating the Project Folder

Create a new project folder with the name *c07_tut1* at *\Documents\3dsmax2017* and save the file with the name *c07tut1*as discussed in Tutorial 1 of Chapter 2.

Opening the File

In this section, you will open the file.

1. Open Windows Explorer and then browse to the *c07_3dsmax_2017_tut* folder and copy all jpeg files from this folder to *\Documents\3dsmax2017\c07_tut1\sceneassets\images*.

2. Choose the **Open** button from the **Application** menu; the **Open File** dialog box is displayed. In this dialog box, browse to *\Documents\c07_3dsmax_2017_tut* and then select the **c07_tut2_start.max** file from it. Choose the **Open** button to open the file, refer to Figure 7-17.

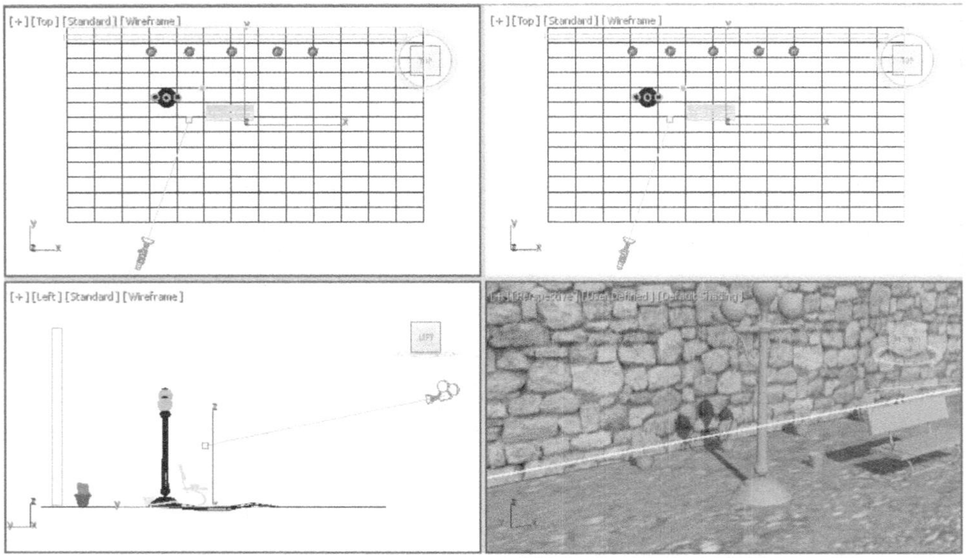

Figure 7-17 *The file displayed*

3. Choose the **Save As** button from the **Application** menu; the **Save File As** dialog box is displayed. Browse to the location *\Documents\3dsmax2017\c07_tut2\scenes*. Save the file with the name *c07tut2.max* at this location.

Creating Material for Bench

In this section, you will create material for bench using the **Standard** and **Architectural** materials.

1. Select *bench* from the Scene Explorer. Next, choose **Group > Open** from the menu bar; the *bench* group is opened.

2. Press the M key; the **Material Editor** dialog box is displayed. Select unused sample slot from the **Material Editor** dialog box. In the **Material Name** drop-down list, enter **wood**.

3. In the **Blinn Basic Parameters** rollout, choose the **Diffuse** map button located next to the **Diffuse** button; the **Material/Map Browser** dialog box is displayed. Choose the **Bitmap** map from **Maps > General** rollout and choose the **OK** button; the **Select Bitmap Image File** dialog box is displayed. As the project folder is already set, the *images* folder is displayed in the **Look in** drop-down list of this dialog box. Select the **wood.jpg** file and then choose the **Open** button; the image is displayed in the selected sample slot.

4. Select *back* from the *bench* group. Next, choose the **Assign Material to Selection** and **Show Shaded Material in Viewport** buttons from the **Material Editor** dialog box; the *wood* material is assigned to *back,* as shown in Figure 7-18.

You will notice that the *wood* material is not displayed properly on *back*. To properly display this material, you need to follow the steps given next.

5. Make sure *back* is selected. Next, select the **UVW Map** modifier from the **Modifier List** drop-down list of the **Command Panel**; the modifier is displayed in the modifier stack and rollouts are displayed below the modifier stack.

6. In the **Parameters** rollout, select the **Box** radio button from the **Mapping** area. Next, select the **X** radio button and then choose the **Fit** button from the **Alignment** area; the texture is properly applied on *back*, as shown in Figure 7-19.

Figure 7-18 *The wood material assigned to back*

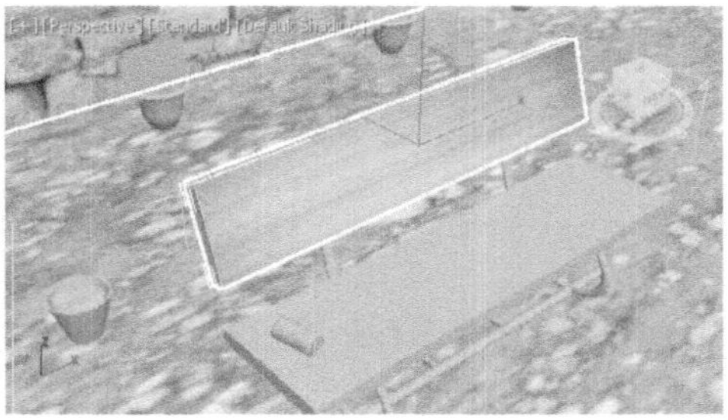

Figure 7-19 *The wood material displayed after using the* ***UVW Map*** *modifier*

7. Select *seat* from the Scene Explorer. In the **Material Editor** dialog box, make sure that *wood* material is selected. Next, choose the **Assign Material to Selection** button from the **Material Editor** dialog box; the material is assigned to *seat*.

8. Repeat the procedure in steps 5 and 6 to apply the **UVW Map** modifier to *seat*. Figure 7-20 shows the *wood* material assigned after using the **UVW Map** modifier.

9. Select all backsupports and supports of the *bench* group from the Scene Explorer. Next, select unused sample slot from the **Material Editor** dialog box. In the **Material Name** drop-down list, enter **black metal**.

10. Choose the **Standard** button on the right of the **Material Name** drop-down list; the **Material/ Map Browser** dialog box is displayed. Select the **Architectural** material from the **Materials > General** rollout and choose **OK**; the **Standard** material changes to **Architectural** material in the **Material Editor** dialog box.

11. In the **Material Editor** dialog box, select **Metal - Flat** from the drop-down list in the **Templates** rollout.

12. In the **Physical Qualities** rollout, choose the **Diffuse Color** color swatch; the **Color Selector: Diffuse** dialog box is displayed. In this dialog box, set the black color and choose **OK**.

13. Choose the **Assign Material to Selection** and **Show Shaded Material in Viewport** buttons from the **Material Editor** dialog box; the *black metal* material is assigned to all supports in the *bench* group.

14. Render the Perspective viewport; the rendered image is displayed, refer to Figure 7-21.

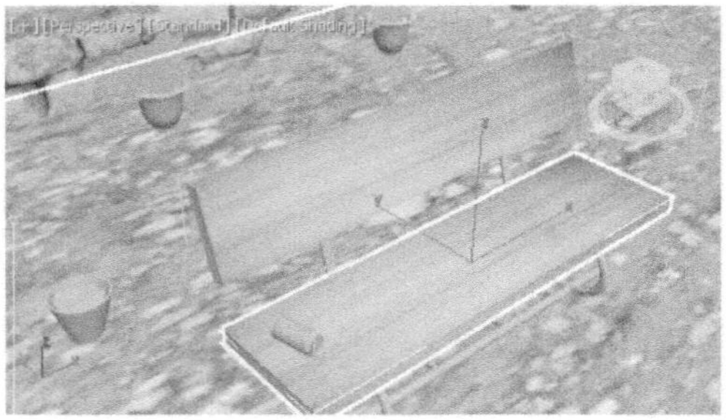

Figure 7-20 *The wood material displayed after using the **UVW Map** modifier*

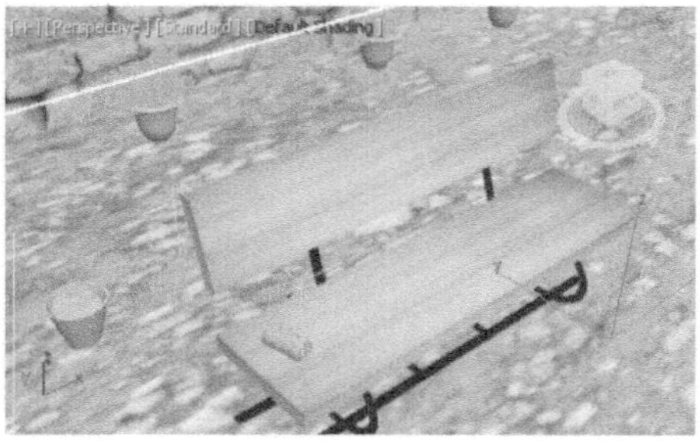

Figure 7-21 *The rendered image*

15. Select *bench* from the Scene Explorer. Next, choose **Group > Close** from the menu bar to close the *bench* group.

Creating Material for Street Lamp

In this section, you will create material for street lamp using the **Standard** and **Architectural** materials.

1. Select *street lamp* from the Scene Explorer. Next, choose **Group > Open** from the menu bar; the *street lamp* group is opened.

2. Select the *black metal* material from the **Material Editor** dialog box and drag it to unused sample slot. Next, change the name in the **Material Name** drop-down list to **green metal**.

3. In the **Physical Qualities** rollout, choose the **Diffuse Color** color swatch; the **Color Selector: Diffuse** dialog box is displayed. In this dialog box, set the values as follows and choose **OK**.

   Red: **16** Green: **28** Blue: **7**

4. Select all parts of *street lamp* except *dome01, dome02,* and *dome03*. Next, choose the **Assign Material to Selection** and **Show Shaded Material in Viewport** buttons from the **Material Editor** dialog box; the *green metal* material is assigned to selected parts of the *street lamp* group.

   Next, you will create material for domes.

5. Select next unused sample slot from the **Material Editor** dialog box. In the **Material Name** drop-down list, enter **dome**.

6. In the **Blinn Basic Parameters** rollout, select the check box in the **Self-Illumination** area; the **Color** spinner is changed to the **Color** color swatch. Choose the **Color** color swatch; the **Self-Illum Color** dialog box is displayed. Enter **198** in the **Red**, **Green**, and **Blue** spinners and choose **OK**.

7. Enter **75** in the **Opacity** spinner located below the **Self-Illumination** area.

8. Select *dome01*, *dome02*, and *dome03* from the Scene Explorer. Next, choose the **Assign Material to Selection** and **Show Shaded Material in Viewport** buttons from the **Material Editor** dialog box; the *dome* material is assigned to *dome01*, *dome02*, and *dome03*.

9. Render the Perspective viewport; the rendered image is displayed, refer to Figure 7-22.

Figure 7-22 *The rendered image*

10. Select *street lamp* from the Scene Explorer. Next, choose **Group > Close** from the menu bar to close the *street lamp* group.

Creating Material for Can

In this section, you will create material for *can* using the **Multi/Sub-Object** material. You will also change the default scanline renderer to mental ray renderer to use the material from the **Autodesk Material Library**.

1. Choose the **Render Setup** tool from the **Main Toolbar**; the **Render Setup: Scanline Renderer** dialog box is displayed.

2. In this dialog box, select **NVIDIA mental ray** from the **Renderer** drop-down list; the **Scanline Renderer** is changed to **NVIDIA mental ray** renderer.

3. Select next unused sample slot from the **Material Editor** dialog box. In the **Material Name** drop-down list, enter **canmat**.

4. Choose the **Standard** button on the right of the **Material Name** drop-down list; the **Material/Map Browser** dialog box is displayed. Select the **Multi/Sub-Object** material from the **Materials > General** rollout and choose **OK**; the **Replace Material** dialog box is displayed. In this dialog box, select the **Discard old material?** radio button and choose **OK**. The **Standard** material is replaced by the **Multi/Sub-Object** material in the **Material Editor** dialog box.

5. In the **Multi/Sub-Object Basic Parameters** rollout, choose the **Set Number** button; the **Set Number of Materials** dialog box is displayed. In this dialog box, enter **2** in the **Number of**

Materials spinner and choose **OK**; sub-materials in the **Multi/Sub-Object Basic Parameters** rollout are reduced to 2.

6. Enter **body** in the **Name** text box of first sub-material and choose the **None** button at the right of this text box; the **Material/Map Browser** dialog box is displayed. In this dialog box, select the **Standard** material from the **Materials > Scanline** rollout and choose **OK**.

7. Select **Metal** from the drop-down list in the **Shader basic Parameters** rollout.

8. Choose the **Diffuse** map button from the **Metal Basic Parameters** rollout of the **Material Editor** dialog box; the **Material/Map Browser** dialog box is displayed. Choose the **Bitmap** map from the **Maps > General** rollout and choose the **OK** button; the **Select Bitmap Image File** dialog box is displayed. As the project folder is already set, the *images* folder is displayed in the **Look in** drop-down list of this dialog box. Select the **pepsi.jpg** file and then choose the **Open** button; the image is displayed in the selected sample slot.

9. Choose the **Go to Parent** tool. Next, enter **74** and **56** in the **Specular Level** and **Glossinesss** spinner of the **Specular Highlights** area, respectively. Next, choose the **Show Shaded Material in Viewport** button.

10. Choose the **Go to Parent** tool again. Next, enter **cap** in the **Name** text box of second sub-material and choose the **None** button at the right of this text box; the **Material/Map Browser** dialog box is displayed. In this dialog box, select **Aluminium - Polished** from the **Autodesk Material Library > Metal > Aluminium** rollout and choose **OK**; the **Autodesk Metal** material is displayed in the **Material Editor** dialog box.

11. In the **Material Editor** dialog box, select **Semi-polished** from the **Finish** drop-down list of the **Metal** rollout. Next, choose the **Go to Parent** tool to return to parent level. Next, choose the **Show Shaded Material in Viewport** button.

12. Select *can* from the **Scene Explorer** and choose the **Modify** tab from the **Command Panel**. Next, click on the arrow sign on the left of the **Editable Poly** in the modifier stack.

13. Select the **Polygon** sub-object level; it gets activated and turns yellow.

14. Click on the cap of *can*; a polygon is selected, as shown in Figure 7-23. Next, enter **2** in the **Set ID** spinner of the **Polygon: Material IDs** area, if not already set.

15. Click on the cap of *can* again and then press CTRL+i; rest of the polygons of *can* are selected, as shown in Figure 7-24. Next, enter **1** in the **Set ID** spinner in the **Polygon: Material IDs** area, if not already set.

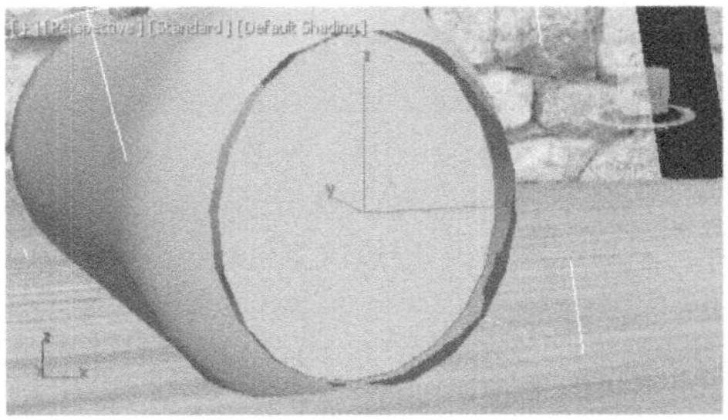

Figure 7-23 Selected polygon of can

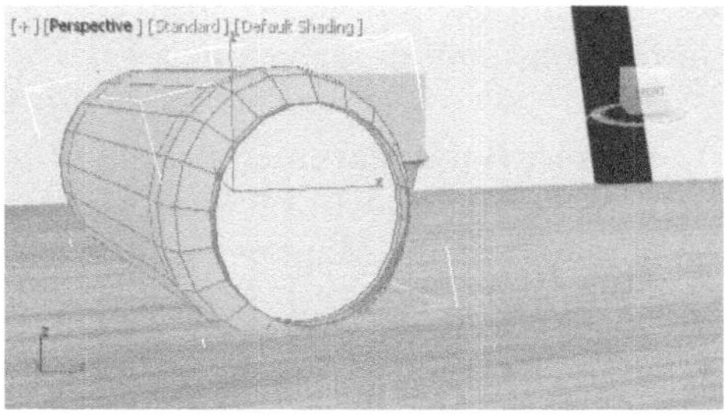

Figure 7-24 Rest of the polygons selected

16. In the modifier stack, select the **Polygon** sub-object level again to deactivate it.

17. Make sure *can* is selected. Next, choose the **Assign Material to Selection** button from the **Material Editor** dialog box; the *canmat* material is assigned to *can*, refer to Figure 7-25.

Creating Material for Dustbin

In this section, you will create material for *dustbin* using the **Standard** material.

1. Select unused sample slot from the **Material Editor** dialog box. In the **Material Name** drop-down list, enter **dustbinmat**.

2. In the **Blinn Basic Parameters** rollout, choose the **Diffuse** map button located next to the **Diffuse** color swatch; the **Material/Map Browser** dialog box is displayed. Choose the **Bitmap** map from the **Maps > General** rollout and choose the **OK** button; the **Select Bitmap Image File** dialog box is displayed. As the project folder is already set, the *images* folder is displayed in the **Look in** drop-down list of this dialog box. Select the **dustbin.jpg** file and then choose the **Open** button; the image is displayed in the selected sample slot.

3. Choose the **Go to Parent** tool. Next, enter **35** and **25** in the **Specular Level** and **Glossinesss** spinners of the **Specular Highlights** area, respectively.

4. Select *dustbin* from the Scene Explorer. Next, assign the *dustbinmat* material to it, as discussed earlier, refer to Figure 7-26.

Figure 7-25 The canmat material displayed

Figure 7-26 The dustbinmat material displayed

Next, you will create material for *garbage*.

5. Select the unused sample slot from the **Material Editor** dialog box. In the **Material Name** drop-down list, enter **garbagemat**.

6. In the **Blinn Basic Parameters** rollout, choose the **Diffuse** map button located next to the **Diffuse** color swatch; the **Material/Map Browser** dialog box is displayed. Choose the **Bitmap** map from the **Maps > General** rollout and choose the **OK** button; the **Select Bitmap Image File** dialog box is displayed. As the project folder is already set, the *images* folder is displayed in the **Look in** drop-down list of this dialog box. Select the **garbage.jpg** file and then choose the **Open** button; the image is displayed in the selected sample slot.

7. Assign the *garbagemat* material to *garbage*, as discussed earlier, refer to Figure 7-27.

Creating Material for Pots and Plants

In this section, you will assign material to pots and plants using the **Standard** material. You will use the **Opacity** map to assign texture to plants.

1. Select *pot_plants* from the Scene Explorer. Next, choose **Group > Open** from the menu bar; the *pot_plants* group is opened.

2. Select the unused sample slot from the **Material Editor** dialog box. In the **Material Name** drop-down list, enter **potmat**.

3. Choose the **Standard** button on the right of the **Material Name** drop-down list; the **Material/ Map Browser** dialog box is displayed. In this dialog box, select **Plaster - Ventetian Rough** from the **Autodesk Material Library > Stucco** rollout and choose **OK**; the **Autodesk Generic** material is displayed in the **Material Editor** dialog box.

4. Choose the color swatch from the **Generic** rollout of the **Material Editor** dialog box; the **Color Selector: Generic_Color** dialog box is displayed. In this dialog box, set the values as follows:

   Red: **0.161** Green: **0.043** Blue: **0**

   Choose **OK** to close the dialog box.

5. Make sure *pot* is selected. Next, select the **UVW Map** modifier from the **Modifier List** drop-down list of the **Command Panel**; the **UVW Map** modifier is displayed in the modifier stack and rollouts are displayed below the modifier stack.

6. In the **Parameters** rollout, select the **Cylindrical** radio button from the **Mapping** area. Next, select the **X** radio button and choose the **Fit** button from the **Alignment** area.

7. Make sure *pot* is selected. Next, assign the *potmat* material to it, as discussed earlier.

   Next, you will create material for *plant1* and *plant2*.

8. Select the unused sample slot from the **Material Editor** dialog box. In the **Material Name** drop-down list, enter **plant**.

9. In the **Blinn Basic Parameters** rollout, choose the **Diffuse** map button located next to the **Diffuse** color swatch; the **Material/Map Browser** dialog box is displayed. Choose the **Bitmap** map from the **Maps > General** rollout and choose the **OK** button; the **Select Bitmap Image File** dialog box is displayed. As the project folder is already set, the *images* folder is displayed in the **Look in** drop-down list of this dialog box. Select the **plant_diff.jpg** file and then choose the **Open** button; the image is displayed in the selected sample slot.

10. Choose the **Go to Parent** tool. Next, choose the **Opacity** map button on the right of the **Opacity** spinner; the **Material/Map Browser** dialog box is displayed. Choose the **Bitmap** map from the **Maps > General** rollout and choose the **OK** button; the **Select Bitmap Image File** dialog box is displayed. As the project folder is already set, the *images* folder is displayed in the **Look in** drop-down list of this dialog box. Select the **plant_op.jpg** file and then choose the **Open** button; the image is displayed in the selected sample slot.

11. Choose the **Go to Parent** tool. Next, assign the *plant* material to *plant1* and *plant2*, as discussed earlier, refer to Figure 7-28.

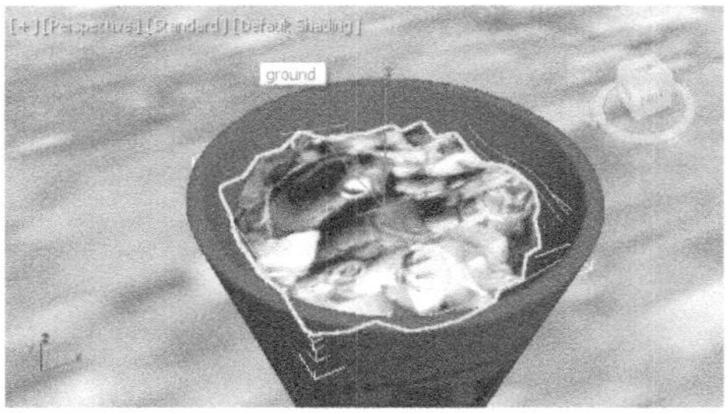

Figure 7-27 *The garbagemat material applied*

Figure 7-28 *The plant material applied*

Note
*You can also insert the **Diffuse** and **Opacity** maps using the **Maps** rollout.*

Next, you will apply material to *soil*.

12. Select the unused sample slot from the **Material Editor** dialog box. In the **Material Name** drop-down list, enter **clay**.

13. In the **Blinn Basic Parameters** rollout, choose the **Diffuse** map button located next to the **Diffuse** color swatch; the **Material/Map Browser** dialog box is displayed. Choose the **Bitmap** map from the **Maps > General** rollout and choose the **OK** button; the **Select Bitmap Image File** dialog box is displayed. As the project folder is already set, the *images* folder is displayed in the **Look in** drop-down list of this dialog box. Select the **clay.jpg** file and then choose the **Open** button; the image is displayed in the selected sample slot.

14. Choose the **Go to Parent** tool. Next, assign the *clay* material to *soil,* as discussed earlier, refer to Figure 7-29.

Figure 7-29 *The* **clay** *material applied to soil*

15. Select *pot_plants* from the Scene Explorer. Next, choose **Group > Close** from the menu bar to close the *pot_plants* group.

16. Similarly, open groups *pot_plants001*, *pot_plants002*, *pot_plants003*, and *pot_plants004*. Next, apply the **UVW Map** modifier to *pot001*, *pot002*, *pot003*, and *pot004*.

17. Assign *pot*, *plant*, and *clay* material to corresponding objects in these groups and then close the groups.

18. Close the **Material Editor** dialog box.

Saving and Rendering the Scene

In this section, you will save the scene and then render it. You can also view the final rendered image of this model by downloading the *c07_3dsmax_2017_rndr.zip* file from *www.cadsofttech.com*. The path of the file is as follows: *Textbooks > Animation and Visual Effects > 3ds Max > Autodesk 3ds Max 2017 for Novices*.

1. Choose **Save** from the **Application** menu.

2. Activate the Perspective viewport. Next, choose the **Render Production** tool from the **Main Toolbar**; the **Rendered Frame** window is displayed. This window shows the final output of the scene, refer to Figure 7-30.

Figure 7-30 The final output of the scene

EXERCISES

The rendered output of the models used in the following exercises can be accessed by downloading the *c07_3dsmax_2017_exr.zip* from *www.cadsofttech.com*.The path of the file is as follows: *Textbooks > Animation and Visual Effects > 3ds Max > Autodesk 3ds Max 2017 for Novices.*

Exercise 1

Create the model of cups and glasses shown in Figure 7-31 using the **Line** tool and the **Lathe** modifier. To create the steel material, you need to use the **Oren-Nayar-Blinn** shader in the **Standard** material and the **Reflection** map in the **Maps** rollout. **(Expected time: 15 min)**

Figure 7-31 The model of cups and glasses

Exercise 2

Create the model of a tea cup in the viewport, as shown in Figure 7-32. Next, create copper material using the **Phong** shader from the **Standard** material and assign it to the object.

(Expected time: 15 min)

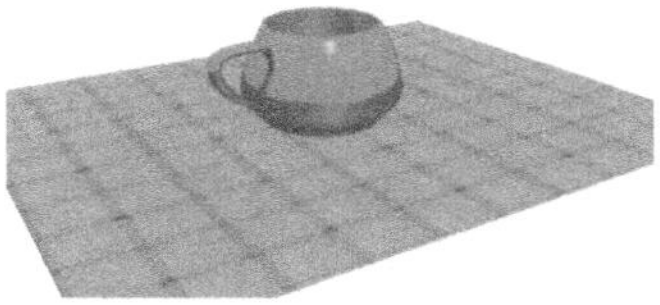

Figure 7-32 *The model of a tea cup*

Chapter 8

Modifying 3D Mesh Objects

Learning Objectives

After completing this chapter, you will be able to:

- *Convert objects into editable mesh*
- *Use sub-object levels*
- *Convert objects into editable poly*

INTRODUCTION

In the previous chapters, you have learned to create different 3D objects and shapes using the tools in the **Create** tab of the **Command Panel**. In this chapter, you will learn to modify the 3D mesh objects at an advanced level.

MODIFYING 3D OBJECTS

To modify 3D objects, select the object and choose the **Modify** tab in the **Command Panel**; the Modify panel will be displayed with various rollouts. In the Modify panel, the default name and color of the object will be displayed at the top. The modifier stack displays the name of the tools that you have used for creating the object, as shown in Figure 8-1.

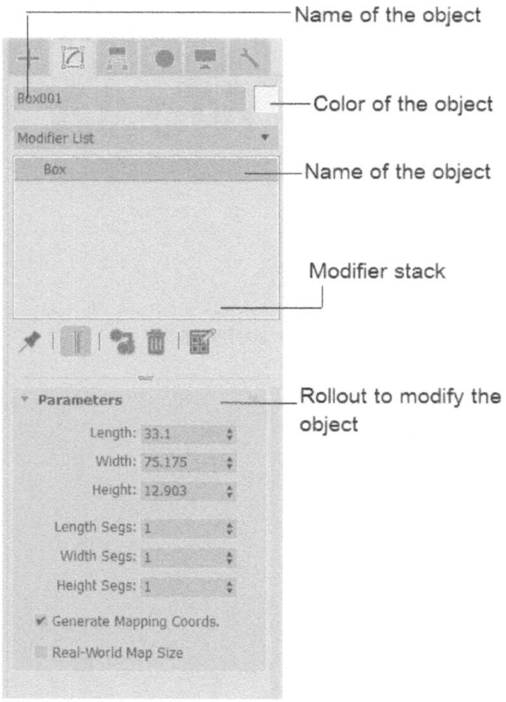

Figure 8-1 *The Modify panel*

Note

The rollouts at the bottom of the Modify panel are displayed based on the type of the object selected.

You can modify the selected object using one of the following methods:

1. By entering new parameters in the rollouts displayed in the Modify panel.

2. By applying a modifier from the **Modifier List** drop-down list.

3. By converting the object into editable mesh or editable poly.

Converting Objects into Editable Mesh

Most of the objects can be modified at advanced level by converting into editable mesh. When open splines such as line and arc are converted into editable mesh, only vertices are available because open splines have no face or edge. When an object is converted into an editable mesh, you get access to its sub-object levels. These sub-object levels help you to select the sub-objects of the object in the viewport. You need to move, rotate, and scale these sub-objects to modify the shape of the object.

You can convert an object into editable mesh using one of the following methods:

1. Create an object and select it in the viewport. Next, right-click on the object; a quad menu will be displayed. Choose the **Convert To:** option; a cascading menu will be displayed, as shown in Figure 8-2. Choose **Convert to Editable Mesh** from the cascading menu; the object will be converted into an editable mesh.

2. Create an object and select it in the viewport. Choose the **Modify** tab in the **Command Panel**; the modifier stack will be displayed in the Modify panel. In the modifier stack, right-click on the name of the tool which has been used to create the object; a shortcut menu will be displayed, as shown in Figure 8-3. Choose the **Editable Mesh** option from the shortcut menu; the object will be converted into an editable mesh.

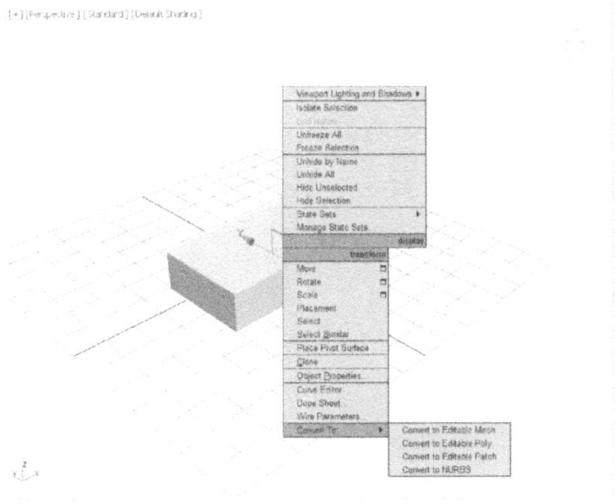

*Figure 8-2 The cascading menu displayed on choosing the **Convert To** option*

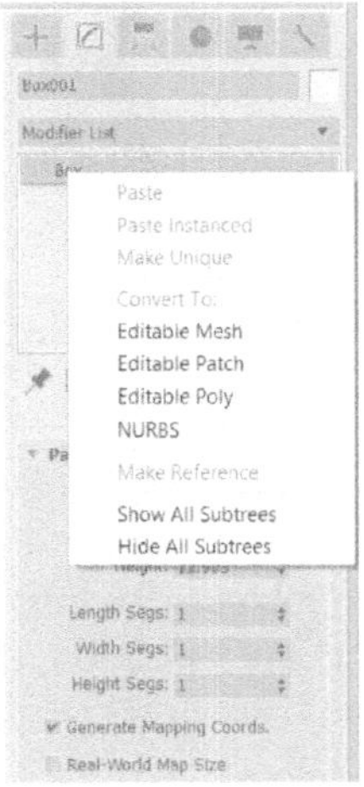

Figure 8-3 The shortcut menu displayed

3. Create an object and select it in the viewport. Choose the **Utilities** tab in the **Command Panel**; the **Utilities** rollout will be displayed. Choose the **Collapse** button in this rollout; the **Collapse** rollout will be displayed. Choose the **Collapse Selected** button in this rollout; the object will be converted into an editable mesh.

Note

*After an object is converted into editable mesh using any of the above method, it cannot be modified using the creation parameters rollouts such as the **Parameters** rollout. So, it is recommended to set the parameters of an object before converting it into an editable mesh.*

You need to use the following method to set the parameters of an object even after converting it into an editable mesh.

4. Create an object and select it in the modifier stack. Next, click on the **Modifier List** drop-down list and select **Edit Mesh** from it; the **Edit Mesh** modifier is applied to the object.

TUTORIAL

Tutorial 1

In this tutorial, you will create the model of an LCD monitor, as shown in Figures 8-4 and 8-5, by modifying 3D mesh objects at advanced level. **(Expected time: 30 min)**

Figure 8-4 *The model of an LCD monitor (view 1)*

Figure 8-5 *The model of an LCD monitor (view 2)*

The following steps are required to complete this tutorial:

a. Create the project folder.
b. Create the front portion of monitor.
c. Create the back portion of monitor.
d. Create speakers.
e. Create buttons.
f. Create the support of monitor.
g. Save and render the scene.

Creating the Project Folder

Create a new project folder with the name *c08_tut2* at *\Documents\3dsmax2017* and then save the file with the name *c08tut2*, as discussed in Tutorial 1 of Chapter 2.

Creating the Front Portion of Monitor

In this section, you will create the front portion of the LCD monitor.

1. Activate the Front viewport and choose **Create > Geometry** in the **Command Panel**. In this panel, by default the **Standard Primitives** option is displayed in the drop-down list below the **Geometry** button. Now, choose the **Box** tool from the **Object Type** rollout.

2. Expand the **Keyboard Entry** rollout and set the following values in their respective spinners:

   Length: **86** Width: **128** Height: **3.0**

3. Choose the **Create** button; a box is displayed in the viewports.

4. In the **Parameters** rollout, set the values as follows:

   Length Segs: **12** Width Segs: **15** Height Segs: **1**

5. Choose the **Zoom Extents All** tool; a box is displayed in all the viewports, as shown in Figure 8-6.

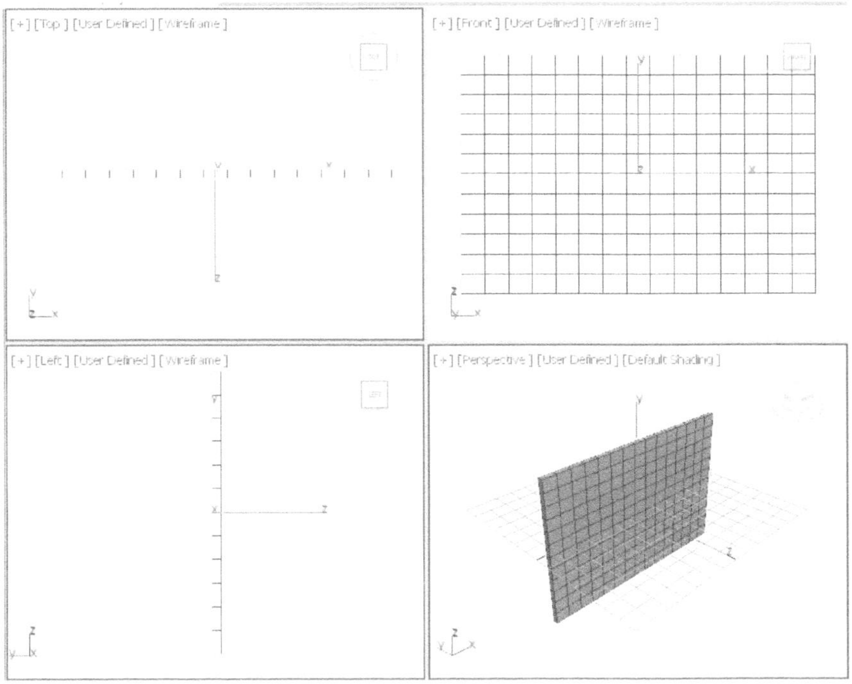

Figure 8-6 *The box for the front portion of the monitor displayed in viewports*

6. In the **Name and Color** rollout, enter **front portion**; the box is named as *front portion*. Also, change the color of *front portion* to white.

Next, you need to convert *front portion* into an editable poly to modify it using sub-object levels.

7. Select *front portion* in any viewport and right-click on it; a quad menu is displayed. Choose the **Convert To:** option; a cascading menu is displayed. Choose **Convert to Editable Poly**; *front portion* is converted into an editable poly.

   When you convert *front portion* into an editable poly, the name of the **Box** tool in the modifier stack gets replaced with **Editable Poly**. Next, you need to modify *front portion* using sub-object levels in the modifier stack to give it a shape similar to the front portion of the monitor.

8. Activate the Front viewport and choose the **Maximize Viewport Toggle** tool to maximize the viewport.

9. Make sure that *front portion* is selected in the viewport. In the modifier stack, click on the arrow sign on the left of the **Editable Poly** to view all the sub-object levels.

10. Select the **Polygon** sub-object; it turns yellow and gets activated.

11. In the **Selection** rollout, select the **Ignore Backfacing** check box, refer to Figure 8-7.

12. In the Front viewport, hold the CTRL key and select all polygons of *front portion*, except the outer rows, refer to Figure 8-8.

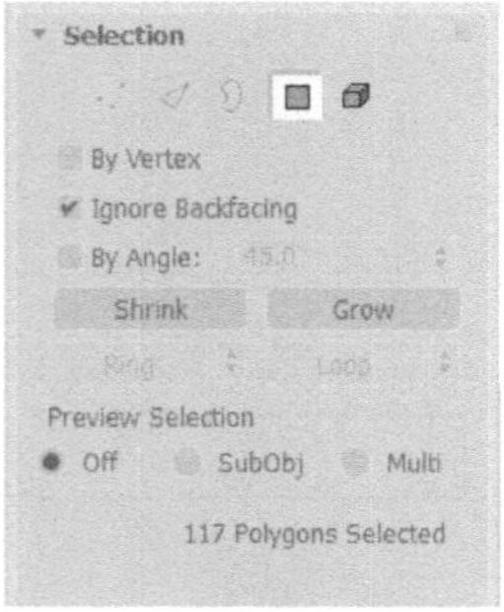

***Figure 8-7** The **Ignore Backfacing** check box selected*

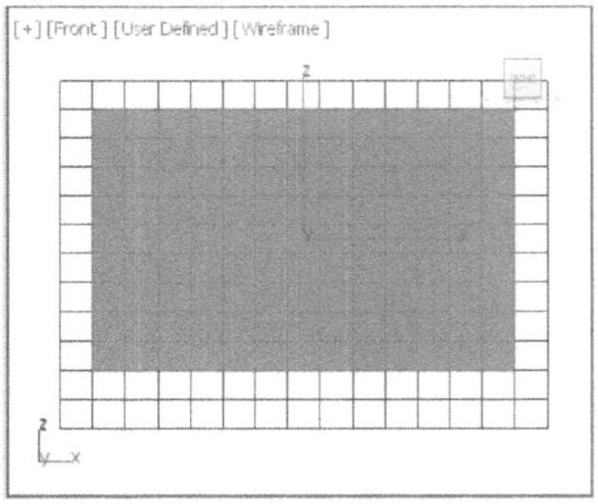

***Figure 8-8** All polygons selected in the Front viewport*

Note

*You can also use the **Grow** button in the **Selection** rollout to increase the selection area of polygons in the viewport.*

13. In the **Edit Polygons** rollout, choose the **Settings** button on the right side of the **Extrude** button to extrude the selected polygons; the **Extrude Polygons** caddy control is displayed in the viewport, as shown in Figure 8-9. In the **Height** spinner, set the value **-3** for negative extrusion and choose **OK** twice to close the caddy control.

14. Click on the **Polygon** sub-object level in the modifier stack to deactivate it. Next, click anywhere in the viewport to deselect *front portion*.

15. Choose the **Maximize Viewport Toggle** tool to view all the four viewports simultaneously. The extrusion effect is displayed in the Perspective viewport, as shown in Figure 8-10.

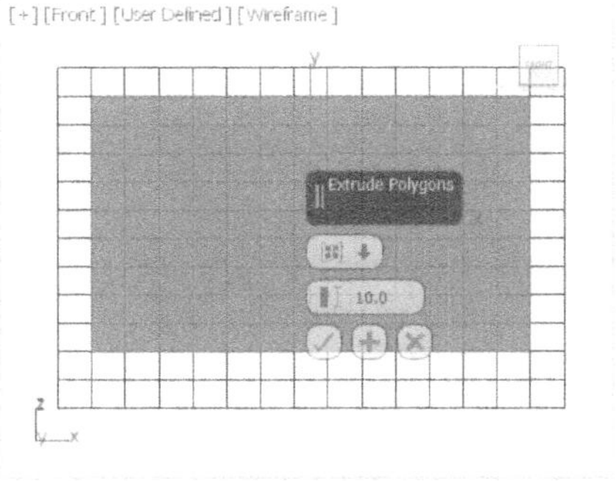

Figure 8-9 The ***Extrude Polygons*** *caddy control*

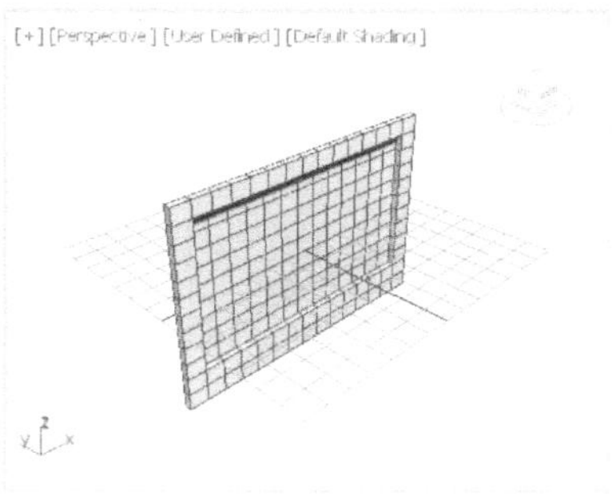

Figure 8-10 The extrusion effect displayed in the Perspective viewport

Now, you need to create a box for the screen of *front portion*.

16. Activate the Front viewport, choose **Create > Geometry** from the **Command Panel**. In this panel, the **Standard Primitives** option is displayed by default in the drop-down list below the **Geometry** button. Now, choose the **Box** tool from the **Object Type** rollout.

17. Expand the **Keyboard Entry** rollout and set the following values in their respective spinners:

    Length: **72** Width: **110.5** Height: **0.5**

18. Choose the **Create** button; a box is displayed in the viewports.

19. In the **Parameters** rollout, set the values as follows:

    Length Segs: **5** Width Segs: **5** Height Segs: **5**

20. In the **Name and Color** rollout, enter **screen**; the box is named as *screen*. Change the color of *screen* to black. Figure 8-11 shows *screen* in the viewports.

21. Group *screen* and *front portion* as discussed earlier and name it as *front part*.

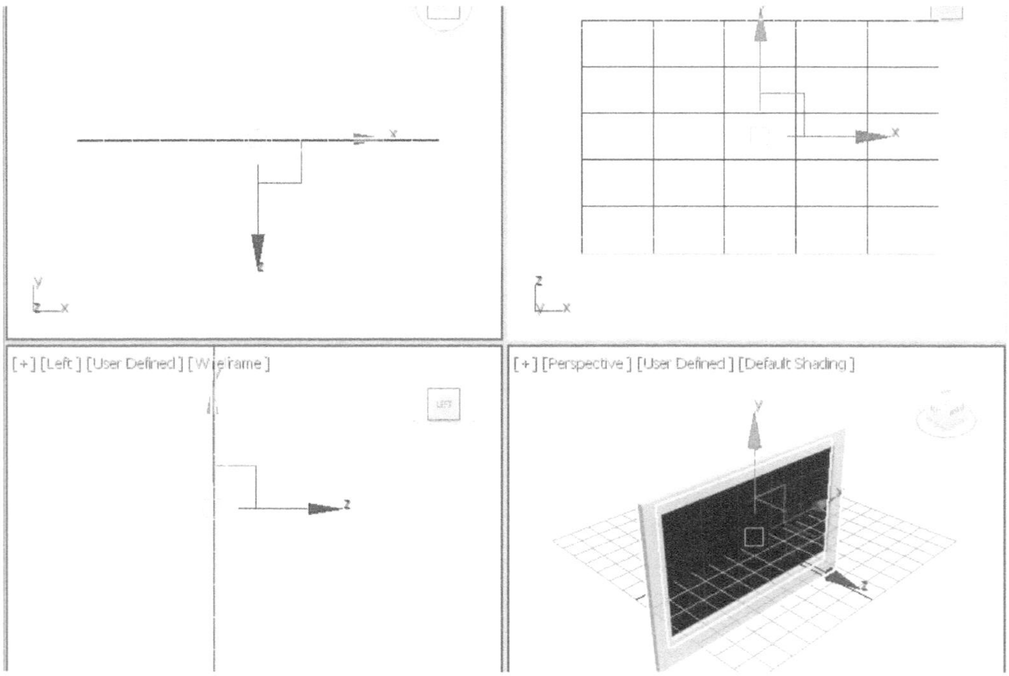

Figure 8-11 The screen geometry in the viewports

Creating the Back Portion of Monitor

In this section, you will create the back portion of the monitor.

1. Activate the Front viewport, choose **Create > Geometry** from the **Command Panel**. Now, choose the **Box** tool from the **Object Type** rollout.

2. Expand the **Keyboard Entry** rollout and set the following values in their respective spinners:

   Length: **86** Width: **128** Height: **5.0**

3. Choose the **Create** button; a box is displayed in viewports.

4. In the **Parameters** rollout, set the values as follows:

   Length Segs: **12** Width Segs: **15** Height Segs: **1**

5. In the **Name and Color** rollout, enter **back portion** and press ENTER; the box is named as *back portion*. Now, change the color of *back portion* to black.

6. Activate the Left viewport and align *back portion* with the back of *front part,* as shown in Figure 8-12.

7. Convert *back portion* into an editable poly to modify it at an advanced level, as discussed earlier.

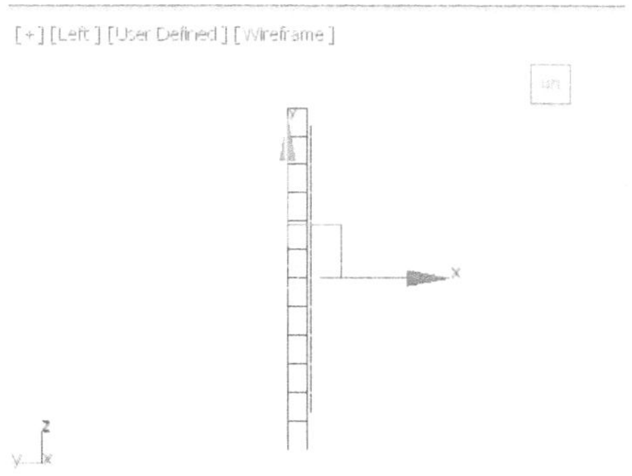

Figure 8-12 *Alignment of the back portion with the front part in the Left viewport*

8. Activate the Front viewport and make sure *back portion* is selected. Now, right-click on *back portion*; a quad menu is displayed. Choose the **Hide Unselected** option from the quad menu; *front part* is hidden in the viewport and only *back portion* is displayed in all the viewports. This will help you to modify *back portion* easily.

   Next, you need to modify the back side of *back portion*. To do so, you need to change the Front viewport to the Back viewport so that you can view the back side of the object.

9. Click on the **Front** label in the Front viewport; a flyout is displayed showing various options to display different views of the viewport, as shown in Figure 8-13. Choose the **Back** option; the Back viewport is displayed.

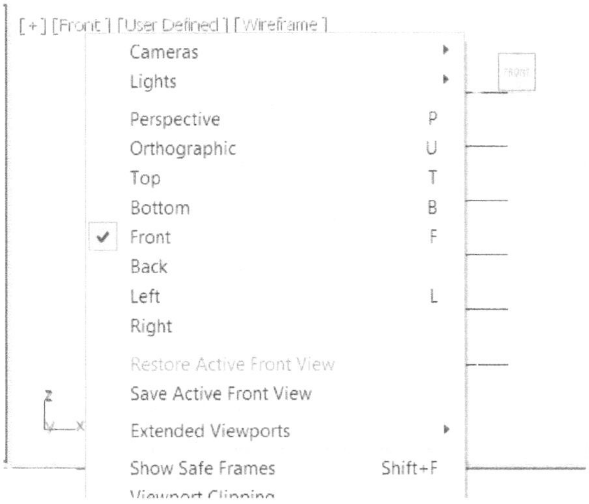

Figure 8-13 *A flyout showing various options to display views*

10. In the modifier stack, click on the arrow sign on the left of **Editable Poly** to view all the sub-object levels.

Tip
*To perform precise modeling at an advanced level, you can invoke the **Maximize Viewport Toggle** tool.*

11. Select the **Polygon** sub-object level; it turns yellow and gets activated.

12. In the Back viewport, hold the CTRL key and select the outer polygons of *back portion*, as shown in Figure 8-14.

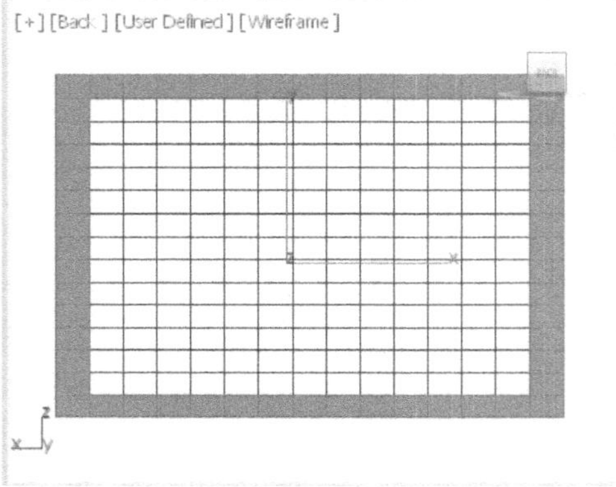

Figure 8-14 *The outer polygons selected in back portion*

13. Right-click on the **Select and Uniform Scale** tool from the **Main Toolbar**; the **Scale Transform Type-In** dialog box is displayed. In the **Offset: Screen** area, set the value **90** in the **%** spinner and press ENTER; the selected polygons are uniformly scaled to 90 percent, as shown in Figure 8-15. Next, close this dialog box.

14. Click anywhere in the viewport to deselect polygons.

15. In the **Selection** rollout, select the **Ignore Backfacing** check box. Next, choose the **Select Object** tool and select the inner polygons by using the rectangular cross selection, as shown in Figure 8-16.

16. Right-click on the **Select and Uniform Scale** tool from the **Main Toolbar**; the **Scale Transform Type-In** dialog box is displayed. In the **Offset: Screen** area, set the value **90** in the **%** spinner and press ENTER; the selected polygons are uniformly scaled to 90 percent, as shown in Figure 8-17. Next, close this dialog box.

17. Activate the Left viewport and make sure that the polygons are still selected. Right-click on the **Select and Move** tool; the **Move Transform Type-In** dialog box is displayed. In the **Offset: Screen** area, enter **-7** in the **X** spinner and press ENTER; the selected polygons move in the horizontal direction, as shown in Figure 8-18. Next, close this dialog box.

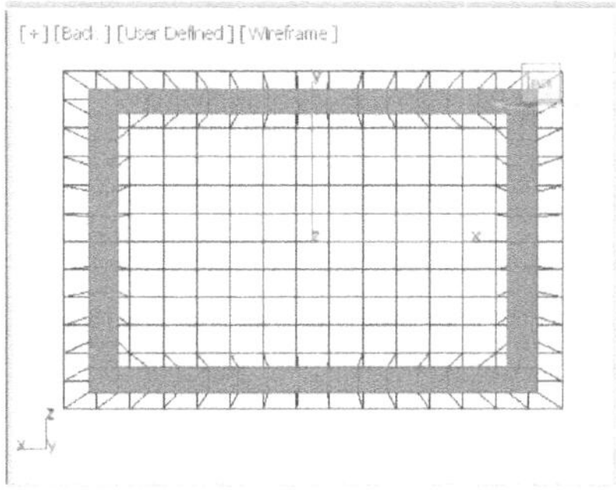

Figure 8-15 The selected polygons uniformly scaled in back portion

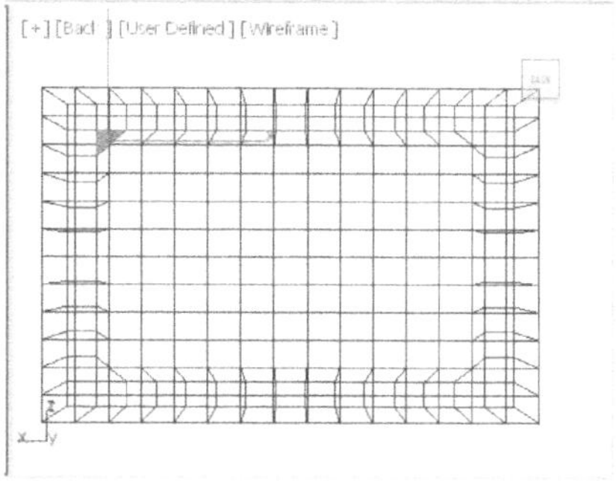

Figure 8-16 Selecting the inner polygons in back portion

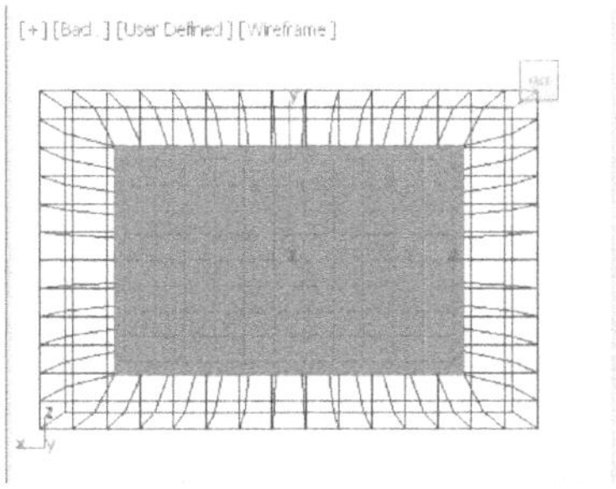

Figure 8-17 The selected polygons uniformly scaled to 90 percent

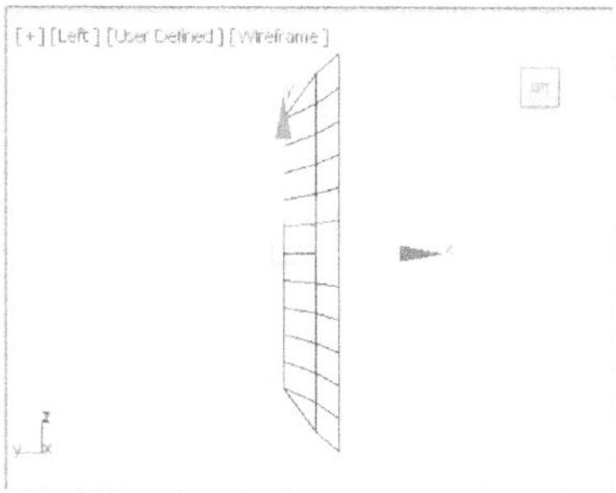

Figure 8-18 Selected polygons dragged horizontally

18. Select the **Polygon** sub-object level in the modifier stack to deactivate it. Next, click anywhere in the viewport to deselect the object in the viewport.

Note

While working in the Back or Left viewport, you can view the output of the modification in the Perspective viewport using the Orbit tool.

19. Change the Back viewport to Front viewport as described earlier. Now, right-click in the viewport; a quad menu is displayed. Choose the **Unhide All** option; all objects are displayed in the viewports.

20. Select *back portion* and *front part* in any viewport and group them as *monitor*.

Creating Speaker for the Monitor

In this section, you will create speaker for the monitor.

1. Activate the Front viewport and choose **Create > Geometry** from the **Command Panel**. By default, the **Standard Primitives** option is displayed in the drop-down list below the **Geometry** button. Now, choose the **Box** tool from the **Object Type** rollout.

2. Expand the **Keyboard Entry** rollout and set the following values in their respective spinners:

   Length: **7.0** Width: **128** Height: **0.2**

3. Choose the **Create** button; a box is displayed in the viewports.

4. In the **Parameters** rollout, make sure the value in the **Length Segs** and **Width Segs** spinners is **15** and in the **Height Segs** spinner is **1**.

5. In the **Name and Color** rollout, enter **speaker**; the box is named as *speaker*. Also, change the color of *speaker* to black.

6. Align *speaker* in viewports with *monitor*, as shown in Figure 8-19.

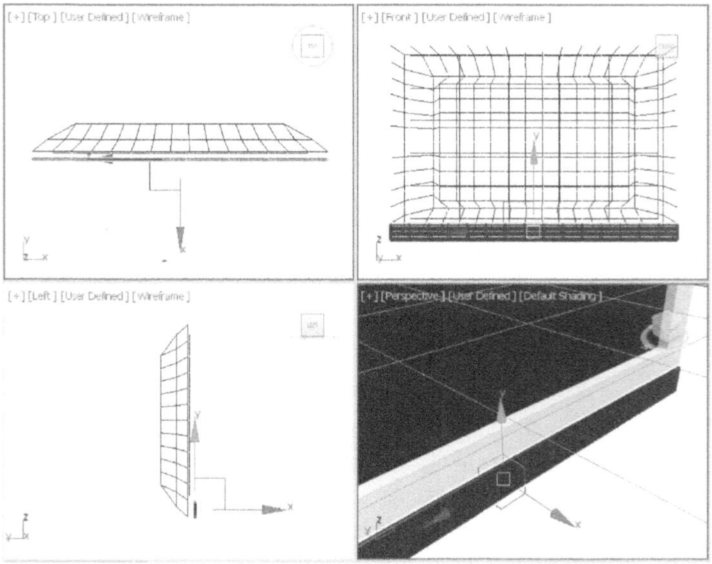

Figure 8-19 *Alignment of the speaker with the monitor in viewports*

You need to apply the **Lattice** modifier to *speaker* from the **Modifier List** drop-down list located at the top of modifier stack.

7. Make sure that *speaker* is selected in the viewport. Next, choose the **Modify** tab in the **Command Panel** and then select the **Lattice** modifier from the **Modifier List** drop-down list; the **Lattice** modifier is displayed in the modifier stack.

8. In the **Parameters** rollout of the **Lattice** modifier, select the **Struts Only from Edges** radio button. In the **Struts** area, set the value **0.5** in the **Radius** spinner. Also, select the **Smooth** check box.

9. Click anywhere in the viewport to deselect *speaker* and choose the **Zoom Extents All** tool to view all objects in all the viewports.

Creating Buttons

In this section, you will create buttons for the monitor by using the **ChamferCyl** tool.

1. Activate the Front viewport and choose **Create > Geometry** in the **Command Panel**. Now, select **Extended Primitives** from the drop-down list below the **Geometry** button and choose the **ChamferCyl** tool from the **Object Type** rollout.

2. Expand the **Keyboard Entry** rollout and set the values as follows:

   Radius: **1.5** Height: **1.0** Fillet: **0.5**

3. Choose the **Create** button; a chamfer cylinder is displayed in the viewports.

4. In the **Name and Color** rollout, enter **b01**; the chamfer cylinder is named as *b01*. Also, change the color of the *b01* to black.

5. Align *b01* in the viewports with *front portion*, as shown in Figure 8-20. You need to choose the **Zoom** or **Zoom Region** tool to align it properly.

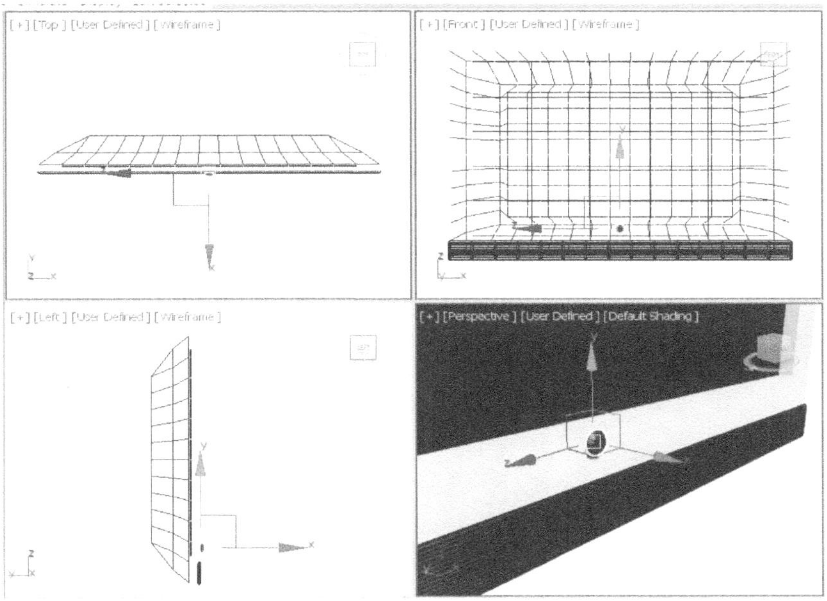

Figure 8-20 *Alignment of b01 with front portion in viewports*

6. In the Front viewport, create five copies of *b01*. They are automatically named as *b002*, *b003*, *b004*, *b005*, and *b006*. Next, align them in the viewports, as shown in Figure 8-21.

7. Select all the objects in the viewport and group them as *LCD Monitor*.

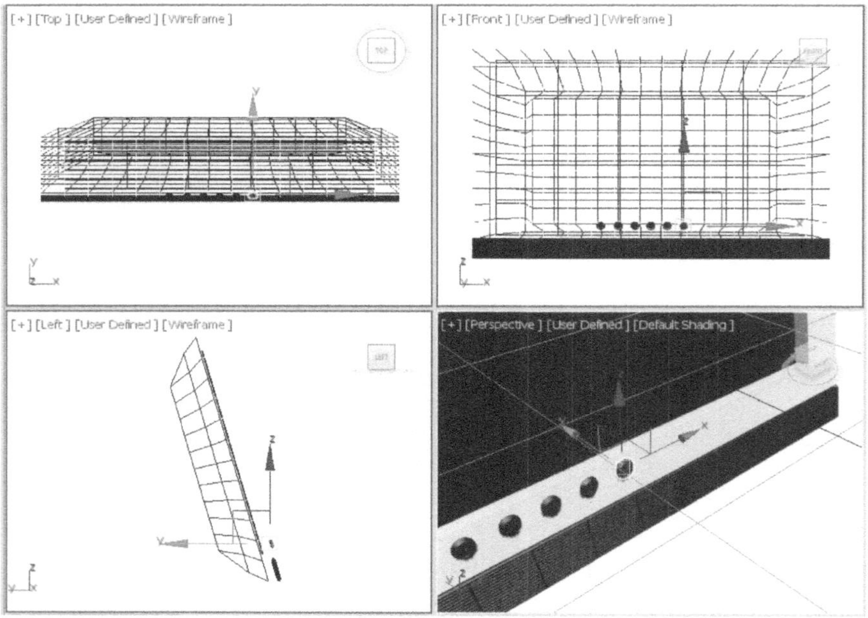

Figure 8-21 *Alignment of b002, b003, b004, b005, and b006 in viewports*

8. Activate the Left viewport and select *LCD Monitor*. Now, right-click on the **Select and Rotate** tool from the **Main Toolbar**; the **Rotate Transform Type-In** dialog box is displayed. In the **Offset: Screen** area, set the value **15** in the **Z** spinner and press the ENTER key.

   Next, close this dialog box; *LCD Monitor* is displayed in the viewports, as shown in Figures 8-22 and 8-23.

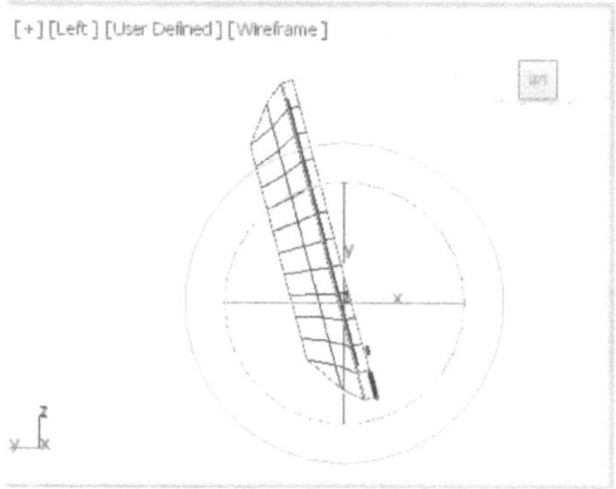

Figure 8-22 *The LCD monitor group displayed after being rotated it in the Left viewport*

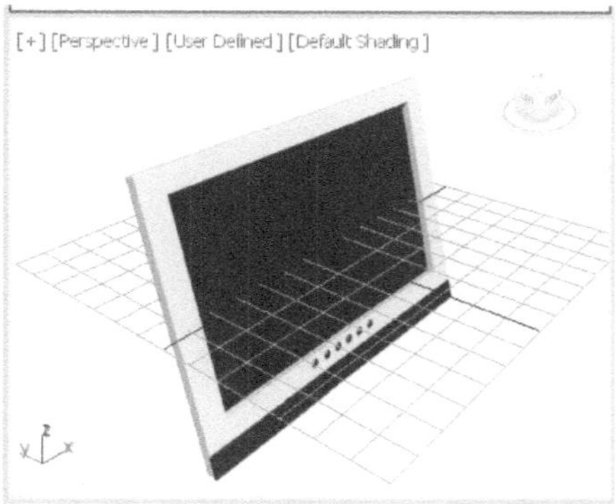

Figure 8-23 *The LCD monitor group displayed in the Perspective viewport*

Creating the Support for the Monitor

In this section, you will create support for the monitor by using the **ChamferCyl** tool

1. Activate the Top viewport and choose **Create > Geometry** in the **Command Panel**. Next, select **Extended Primitives** from the drop-down list below the **Geometry** button, and then choose the **ChamferCyl** tool from the **Object Type** rollout.

2. Expand the **Keyboard Entry** rollout and set the following values in their respective spinners:

   Radius: **23.698** Height: **6.091** Fillet: **2.284**

3. Choose the **Create** button from the **Command Panel**; a chamfer cylinder is displayed in viewports.

4. In the **Parameters** rollout, set the values as follows:

   Fillet Segs: **5** Sides: **18** Cap Segs: **4**

5. In the **Name and Color** rollout, enter **support**; the chamfer cylinder is named as *support*. Also, use the color swatch to change the color of *support* using the following values:

   Red: **115** Green: **115** Blue: **115**

6. In the Left viewport, move the cursor over the Y axis and move *support* below *LCD monitor*. Now move *support* in the horizontal direction, as shown in Figure 8-24.

7. Make sure that *support* is selected. Next, choose the **Zoom Extents All Selected** tool; *support* is zoomed in the viewports, as shown in Figure 8-25.

8. Convert *support* into an editable poly to modify it to an advanced level.

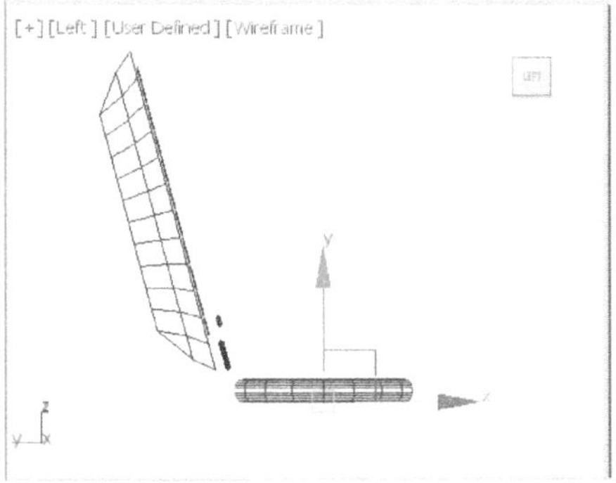

***Figure 8-24** The support geometry moved in the Left viewport*

9. In the modifier stack, click on the plus sign on the left of **Editable Poly** to view all the sub-object levels.

10. Select the **Polygon** sub-object level; it turns yellow and gets activated.

11. Press and hold the CTRL key and select the inner polygons of *support* in the Top viewport, as shown in Figure 8-26.

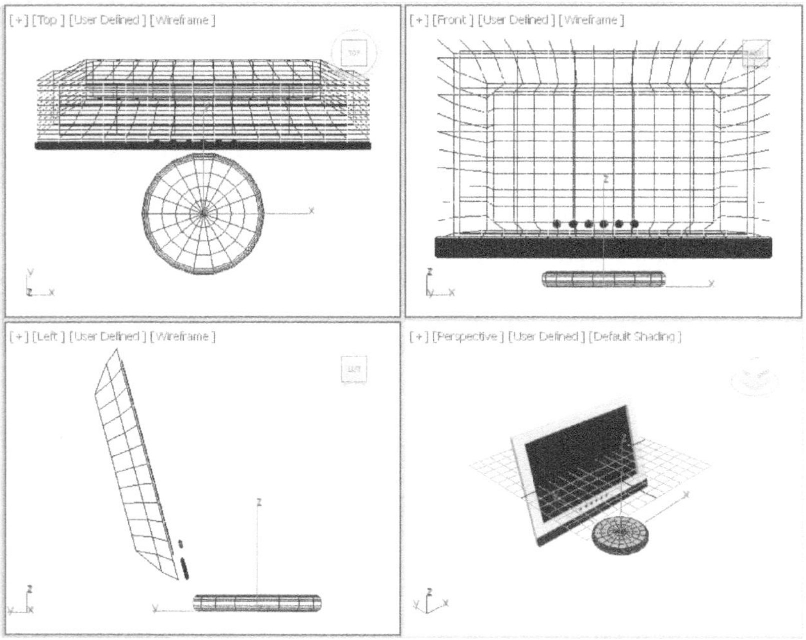

Figure 8-25 *The support geometry zoomed in viewports*

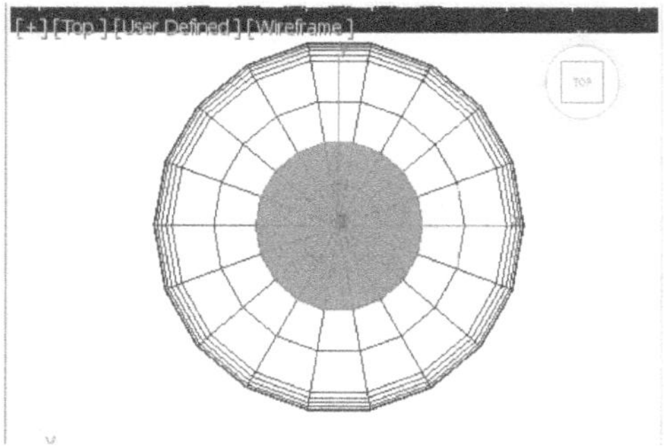

Figure 8-26 *The inner polygons of support selected*

12. In the **Edit Polygons** rollout, choose the **Settings** button on the right of the **Extrude** button; the **Extrude Polygons** caddy control is displayed in the viewport. Set the value **2.0** in the **Height** spinner, refer to Figure 8-27, and choose the **OK** button twice to close this caddy control.

13. Choose the **Settings** button on the right of the **Bevel** button; the **Bevel** caddy control is displayed in the viewport. Set the value **1.0** in the **Height** spinner, refer to Figure 8-28, press ENTER, and choose **OK** twice to close the caddy control; the support is displayed in viewport. Figures 8-29 and 8-30 shows the support displayed in the Top and Perspective viewports, respectively.

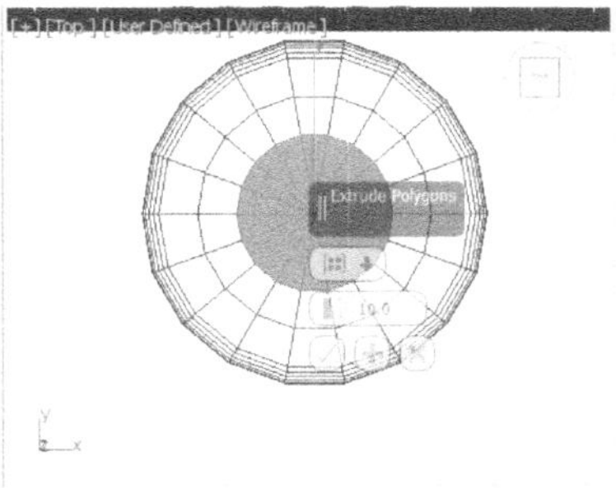

*Figure 8-27 The **Extrude Polygons** caddy control*

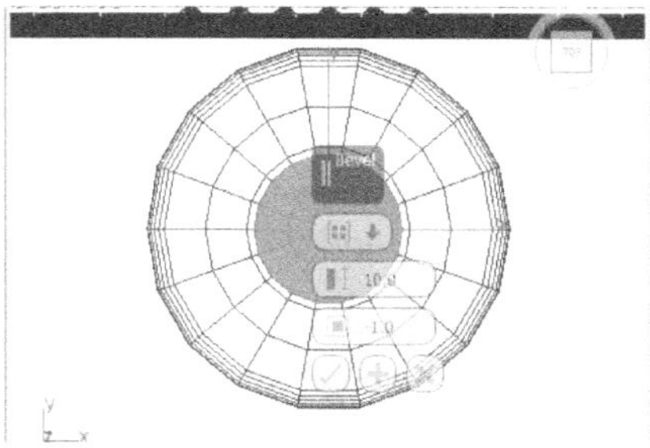

*Figure 8-28 The value set in the **Height** spinner of the **Bevel** caddy control*

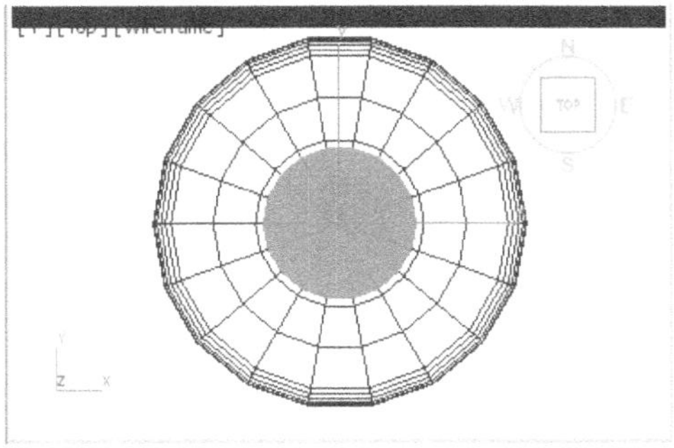

Figure 8-29 The selected polygons in the Top viewport after extruding and beveling

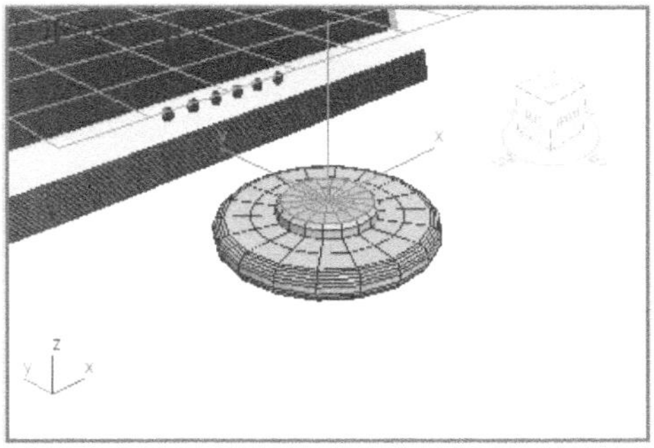

Figure 8-30 The selected polygons in the Perspective viewport after extruding and beveling

14. Select the **Polygon** sub-object level in the modifier stack to deselect it and then click anywhere in the viewport to deselect the object.

15. Choose the **Zoom Extents All** tool and align *support* in the viewports, as shown in Figure 8-31.

16. Select *support* and choose the **Zoom Extents All Selected** tool. Now, select the **Polygon** sub-object level in the modifier stack; all the selected polygons are displayed.

17. Activate the Perspective viewport and set the view using the **Zoom**, **Select and Move** and **Orbit** tools, as shown in Figure 8-32. Next by holding the ALT key, deselect the polygons except the five polygons on the back, refer to Figure 8-33.

18. Select the **Polygon** sub-object level in the modifier stack to deactivate it and then click anywhere in the viewport to deselect the object.

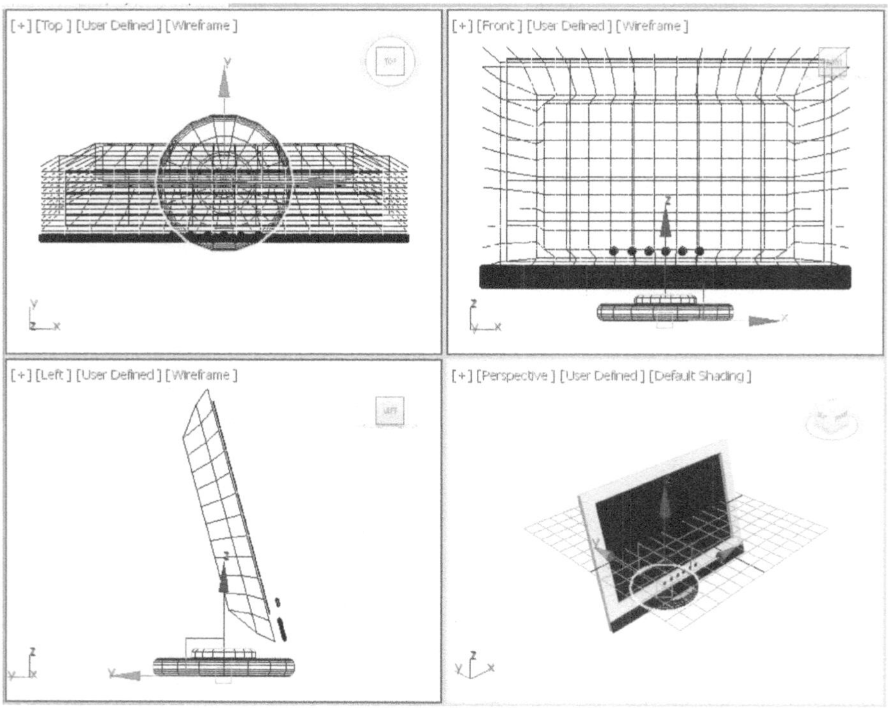

Figure 8-31 *Alignment of support in viewports*

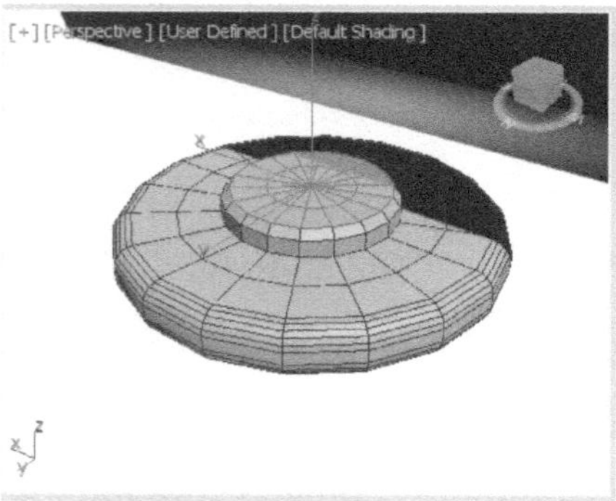

Figure 8-32 *The view of the Perspective viewport after using the **Zoom** and **Orbit** tools*

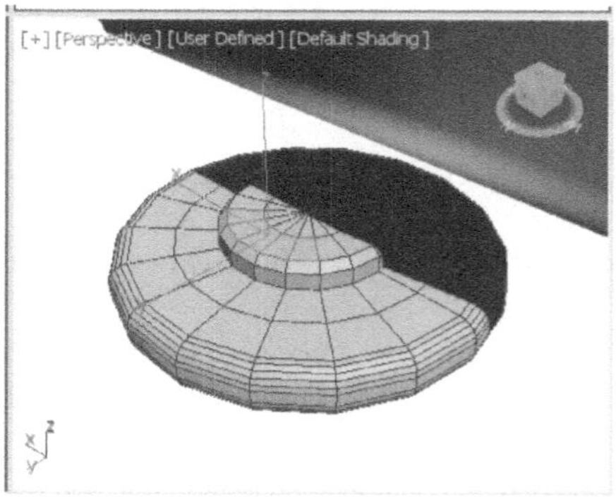

Figure 8-33 *The polygons selected in the Perspective viewport*

19. Choose the **Zoom Extents All** tool and set the view of the Perspective viewport to view the back side of *LCD monitor*, as shown in Figure 8-34.

20. Activate the Left viewport and select *support* from it. Next, select the **Polygon** sub-object level; all the selected polygons are highlighted.

21. In the **Edit Polygons** rollout, choose the **Settings** button on the right of the **Extrude** button; the **Extrude Polygons** caddy control is displayed. Set the value **29** in the **Height** spinner, press ENTER and choose **OK** twice to close this caddy control; the selected polygons are extruded and displayed in the Left viewport, as shown in Figure 8-35.

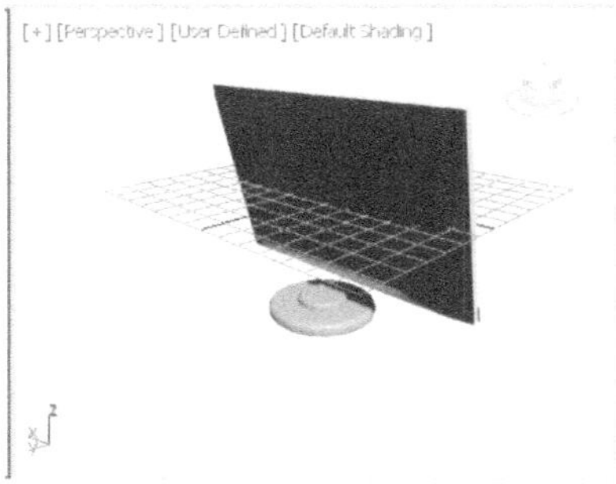

Figure 8-34 *The back side geometry of LCD monitor in the Perspective viewport*

Figure 8-35 *The selected polygons extruded in the Left viewport*

22. In the Left viewport, move the cursor over the horizontal axis and drag the selected polygons horizontally toward the right side until they touch *LCD monitor*, as shown in Figure 8-36.

23. In the **Edit Polygons** rollout, again choose the **Settings** button on the right of the **Extrude** button; the **Extrude Polygons** caddy control is displayed. Set the value **19** in the **Height** spinner and press ENTER. Next, choose **OK** twice to close this caddy control. The *support* is extruded in the Left viewport, as shown in Figure 8-37.

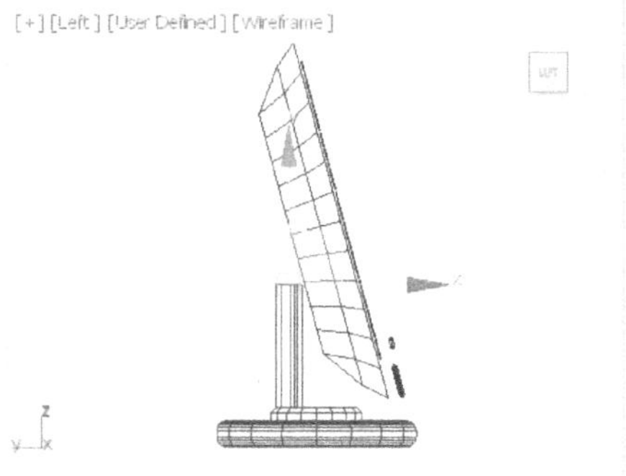

Figure 8-36 *The support geometry in the Left viewport after moving it horizontally*

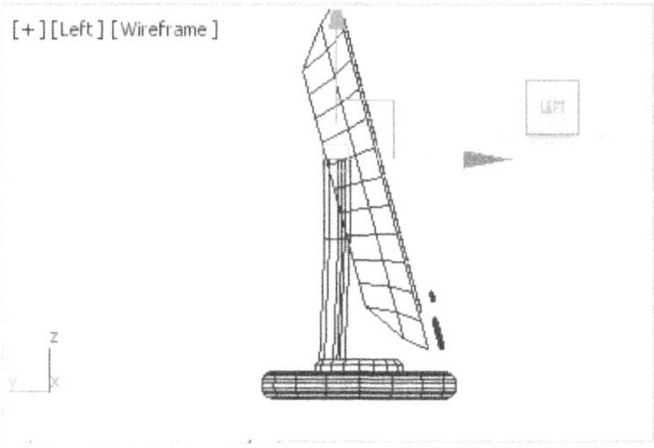

Figure 8-37 *The support geometry in the Left viewport after extrusion*

24. In the Left viewport, move the cursor over the horizontal axis, and drag the selected polygons horizontally toward the left to align them with *LCD monitor*, as shown in Figure 8-38.

25. Select the **Polygon** sub-object level in the modifier stack to deselect it.

Saving and Rendering the Scene

In this section, you will save the scene and then render it. You can also view the final rendered image of this model by downloading the file *c08_3dsmax_2017_rndr.zip* from *www.cadsofttech.com*. The path of the file is as follows: *Textbooks > Animation and Visual Effects > 3ds Max > Autodesk 3ds Max 2017 for Novices*.

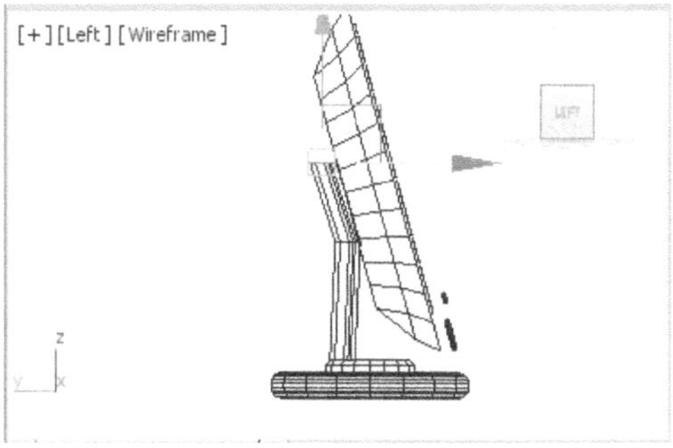

Figure 8-38 *The selected polygons of support in the Left viewport*

1. Choose **Save** from the **Application** menu.

2. Activate the Perspective viewport. Next, choose the **Render Production** tool from the **Main Toolbar**; the **Rendered Frame** window is displayed.

EXERCISE

The rendered output of the model used in the following exercise can be accessed by downloading the *c08_3dsmax_2017_exr.zip* from *www.www.cadsofttech.com*.The path of the file is as follows: *Textbooks > Animation and Visual Effects > 3ds Max > Autodesk 3ds Max 2017 for Novices.*

Exercise 1

Create the model of a bed shown in Figure 8-39 using dimensions of your choice.

(Expected time: 30 min)

Figure 8-39 *The model of a bed*

This page is intentionally left blank

Chapter 9

Graphite Modeling Technique

Learning Objectives

After completing this chapter, you will be able to:

- *Customize the graphite modeling toolset interface*
- *Work with panels in the graphite modeling toolset*

INTRODUCTION

The graphite modeling is a technique in 3ds Max which is used to simplify the modeling process. The tools in the graphite modeling toolset are used for editing mesh and polygonal objects. These tools and their usage are discussed in this chapter.

GRAPHITE MODELING TOOLSET

The graphite modeling toolset is available in the ribbon below the **Main Toolbar**, as shown in Figure 9-1 and is used for polygonal modeling.

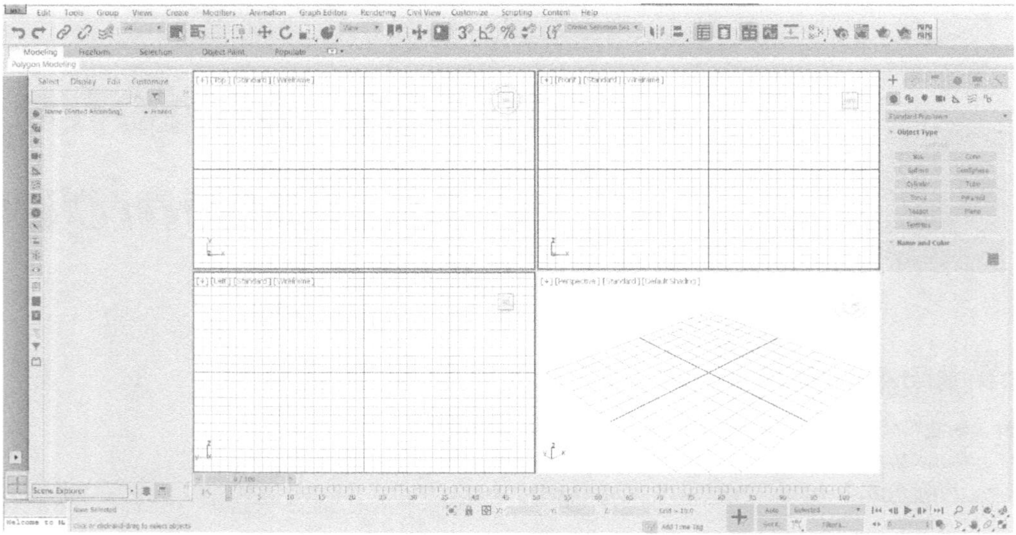

Figure 9-1 *The graphite modeling toolset*

If this toolset is not displayed, right click on the **Main Toolbar**; a shortcut menu will be displayed, as shown in Figure 9-2. Select the **Ribbon** option from the shortcut menu; the graphite modeling toolset will be displayed below the **Main Toolbar**.

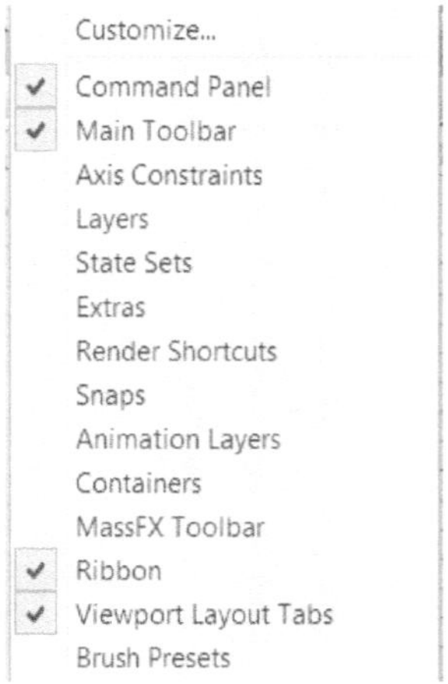

Figure 9-2 *The shortcut menu displayed on right-clicking the* ***Main Toolbar***

Modeling Tab

The **Modeling** tab contains a number of tools that are required for polygon modeling. These tools are organized into various panels.

Polygon Modeling Panel

The tools in the **Polygon Modeling** panel are used to switch between different sub-object levels, convert objects into an editable poly or edit poly objects, and so on. To convert an object into an editable poly, select the object in the viewport and double-click on the **Modeling** tab from the ribbon; the **Polygon Modeling** panel will be displayed, as shown in Figure 9-3. Next, expand the **Polygon Modeling** panel, refer to Figure 9-4. Now, choose the **Convert to Poly** tool from the expanded panel; the selected object will be converted into an editable poly. Also, various panels will be displayed in the **Modeling** tab such as **Edit**, **Geometry (All)**, **Subdivision**, and so on.

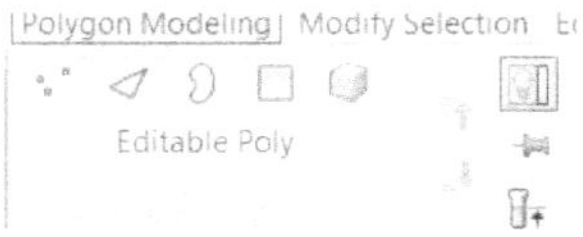

Figure 9-3 The ***Polygon Modeling*** *panel*

Figure 9-4 The ***Polygon Modeling*** *panel expanded*

TUTORIAL

Tutorial 1

In this tutorial, you will create the model of computer table, as shown in Figure 9-5, using the graphite modeling toolset. **(Expected time: 30 min)**

Figure 9-5 The model of a computer table

The following steps are required to complete this tutorial:

a. Create the project folder.
b. Create the tabletop and supports.
c. Create the CPU base.
d. Create top and bottom shelves.
e. Create the keyboard support.
f. Save and render the scene.

Creating the Project Folder

Create a new project folder with the name *c09_tut1* at *\Documents\3dsmax2017* and then save the file with the name *c09tut1*, as discussed in Tutorial 1 of Chapter 2.

Creating the Tabletop and Supports

In this section, you will create the tabletop and supports using the **Box** tool.

1. Activate the Top viewport and choose **Create > geometry** from the **Command Panel**. Now, choose the **Box** tool from the **Object Type** rollout. Next, create a box in the Top viewport.

2. Choose the **Modify** tab from the **Command Panel**. Next, in the **Parameters** rollout, set the parameters as given next:

   Length: **121** Width: **234** Height: **9**

   Length Segs: **10** Width Segs: **24**

3. Rename *box001* as *computer table*. Next, choose the **Zoom Extents All** tool. The *computer table* is displayed in the viewports, as shown in Figure 9-6.

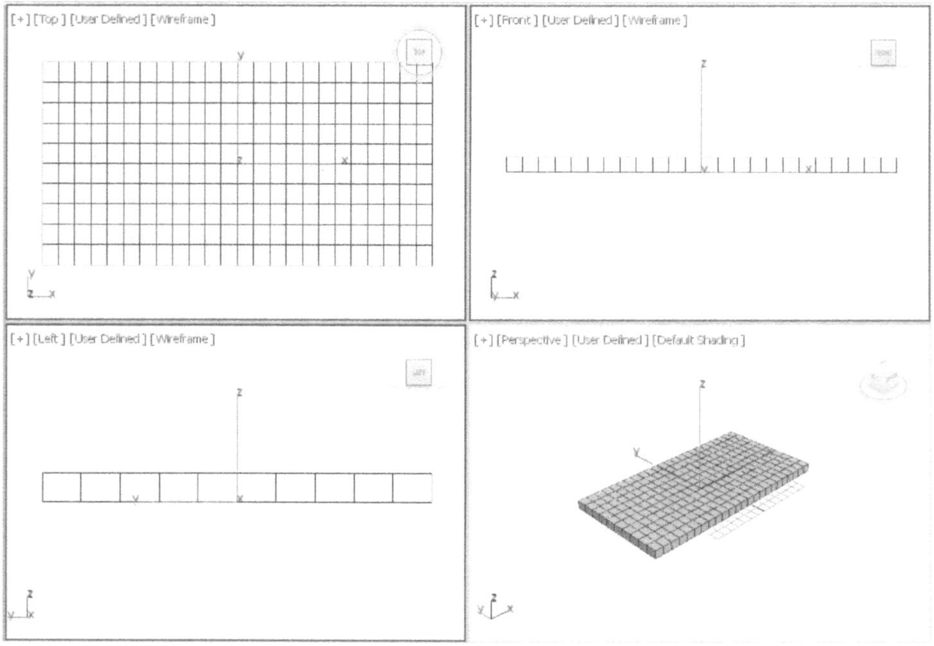

***Figure 9-6** The computer table geometry displayed in the viewports*

4. Choose the **Modeling** tab below the **Main Toolbar**. Next, expand the **Polygon Modeling** panel and then choose the **Convert to Poly** tool from it; *computer table* is converted into an editable poly.

5. Switch from the Top viewport to the Bottom viewport and select the **Polygon** sub-object level. Next, select the polygons of *computer table*, as shown in Figure 9-7.

6. Choose the **Extrude Settings** button from the **Extrude** drop-down in the **Polygons** panel, as shown in Figure 9-8; the **Extrude** caddy control is displayed in the viewport, refer to Figure 9-9.

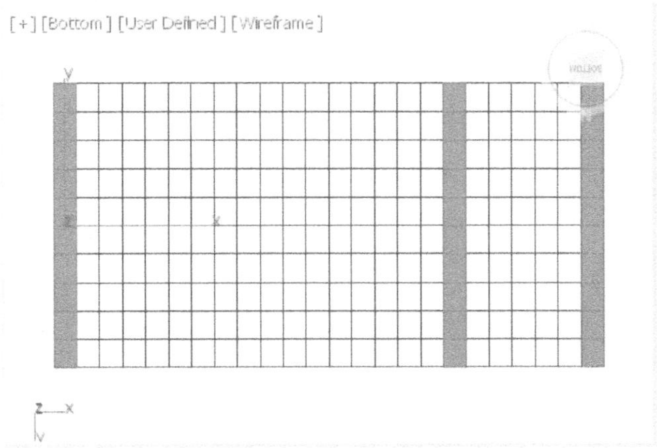

Figure 9-7 *Selected polygons of computer table*

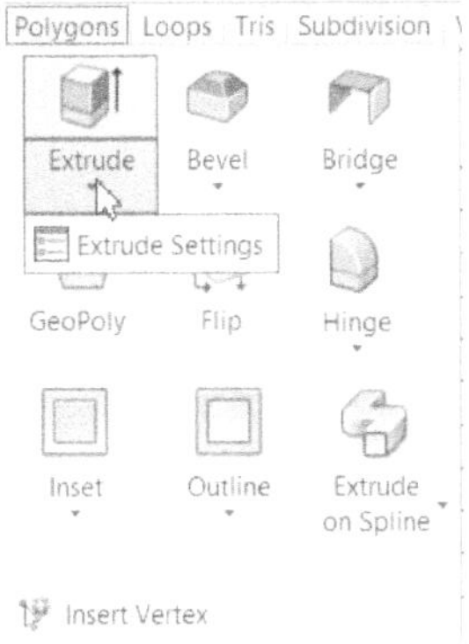

Figure 9-8 *Choosing* ***Extrude Settings*** *from the* ***Extrude*** *drop-down*

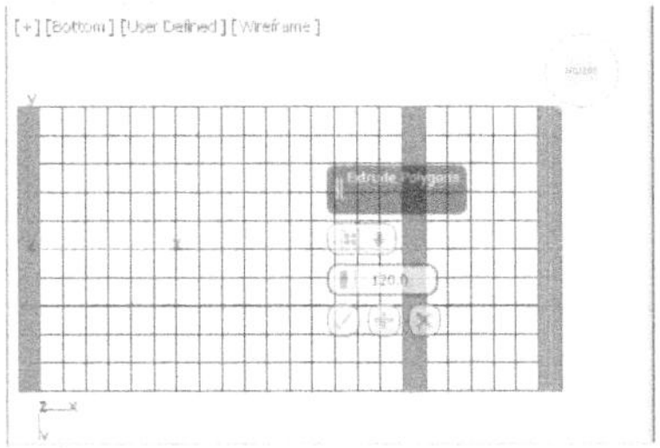

Figure 9-9 *The* ***Extrude*** *caddy control*

7. Set the value **110** in the **Height** spinner and choose the **OK** button; the selected polygons are extruded, as shown in Figure 9-10.

Creating the CPU Base

In this section, you will create the CPU base by extruding the polygons of *computer table.*

1. Switch from the Bottom viewport to the Top viewport. Next, select the **Edge** sub-object level. Activate the Perspective viewport and then select an edge from each extruded part, as shown in Figure 9-11.

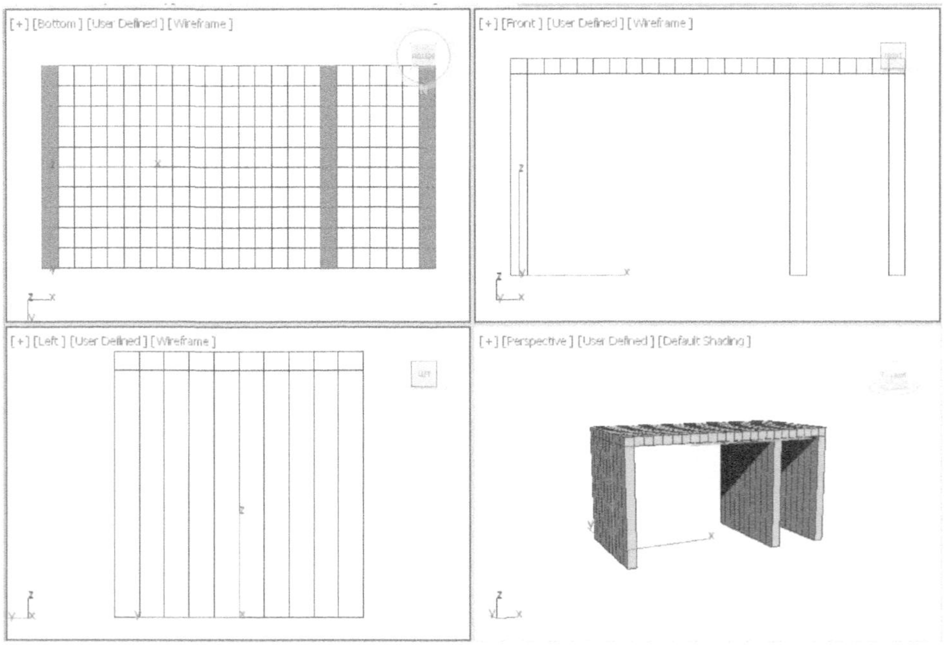

Figure 9-10 *The selected polygons extruded*

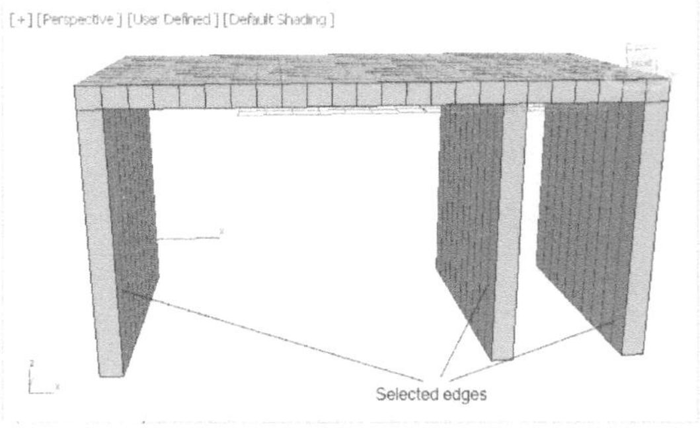

Figure 9-11 *Selected edges of computer table*

2. Choose the **Ring** tool from the **Modify Selection** panel; rings of all the selected edges are selected.

3. Choose the **Connect Settings** button from the **Connect** drop-down in the **Loops** panel; the **Connect** caddy control is displayed in the viewport.

4. In the **Connect** caddy control, set the value **12** in the **Segments** spinner and choose the **OK** button; the edges are added as shown in Figure 9-12.

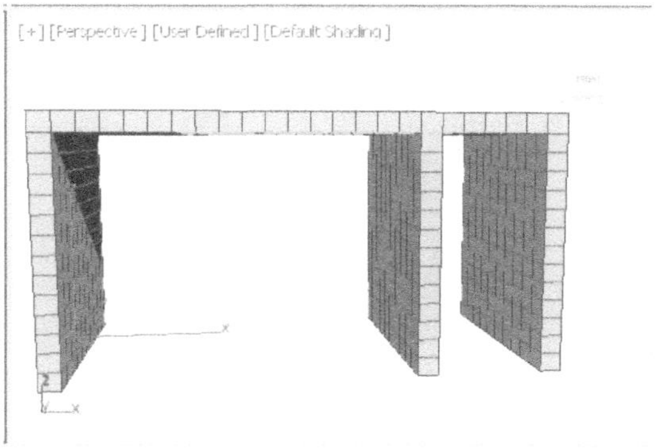

Figure 9-12 *Edges added to the computer table*

5. Select the **Polygon** sub-object level. Next, select the polygons, as shown in Figure 9-13.

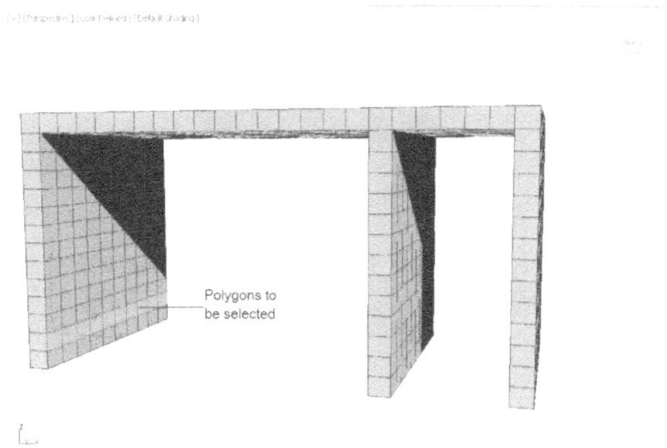

Figure 9-13 *The selected polygons of computer table*

6. Choose the **Extrude Settings** button from the **Extrude** drop-down in the **Polygons** panel; the **Extrude** caddy control is displayed in the viewport. In this caddy control, set the value **55** in the **Height** spinner and choose the **OK** button; the selected polygons are extruded, as shown in Figure 9-14.

7. Select the **Edge** sub-object level and then select an edge from the extruded part, as shown in Figure 9-15.

8. Choose the **Ring** tool from the **Modify Selection** panel; ring of the selected edge is selected.

9. Choose the **Connect Settings** button from the **Connect** drop-down in the **Loops** panel; the **Connect** caddy control is displayed in the viewport.

10. In the **Connect** caddy control, set the value **8** in the **Segments** spinner and choose the **OK** button; the edges are added, as shown in Figure 9-16.

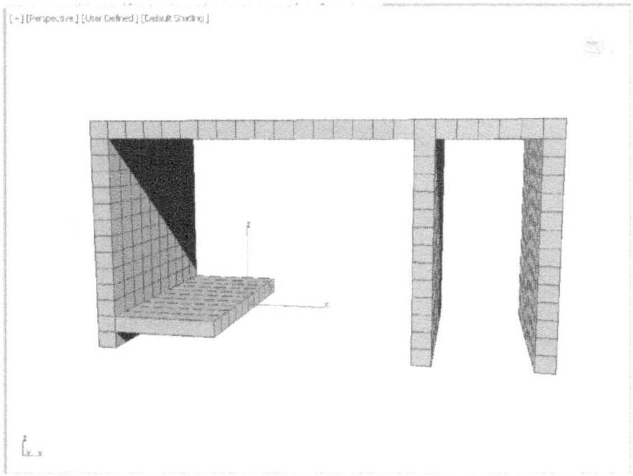

Figure 9-14 *The extruded polygons of computer table*

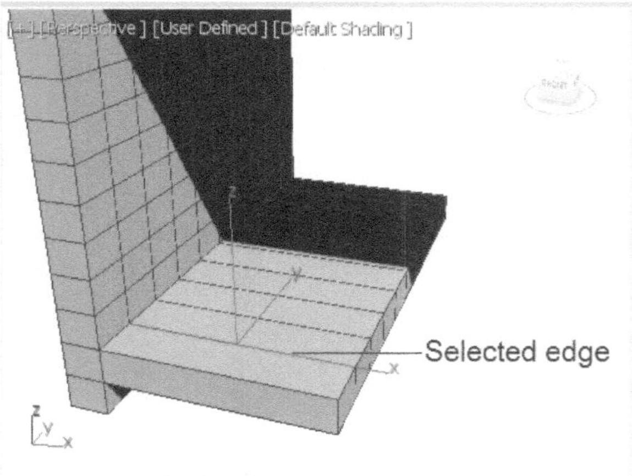

Figure 9-15 *The selected edge of computer table*

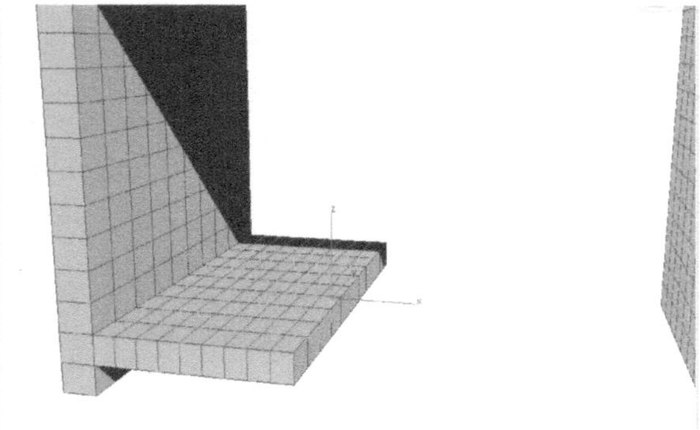

Figure 9-16 *Edges added to the computer table*

11. Select the **Polygon** sub-object level and select the polygons of the extruded part, as shown in Figure 9-17.

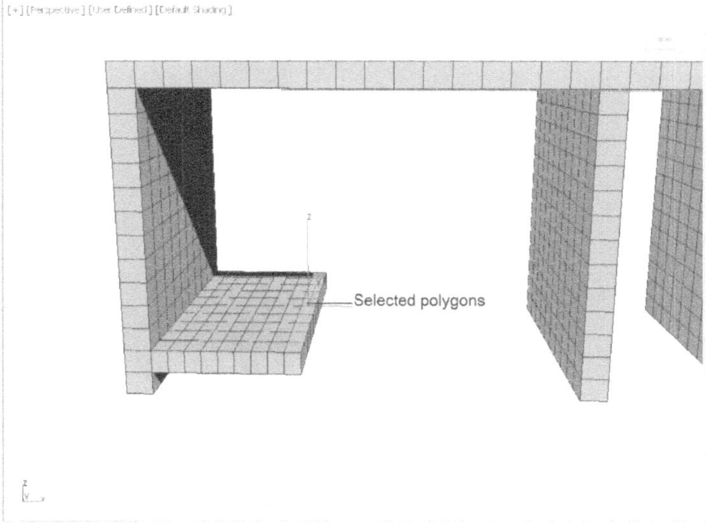

Figure 9-17 *Selected polygons of computer table*

12. Choose the **Extrude Settings** button from the **Extrude** drop-down in the **Polygons** panel; the **Extrude** caddy control is displayed in the viewport. In this caddy control, set the value **30** in the **Height** spinner and choose the **OK** button; the selected polygons are extruded, as shown in Figure 9-18.

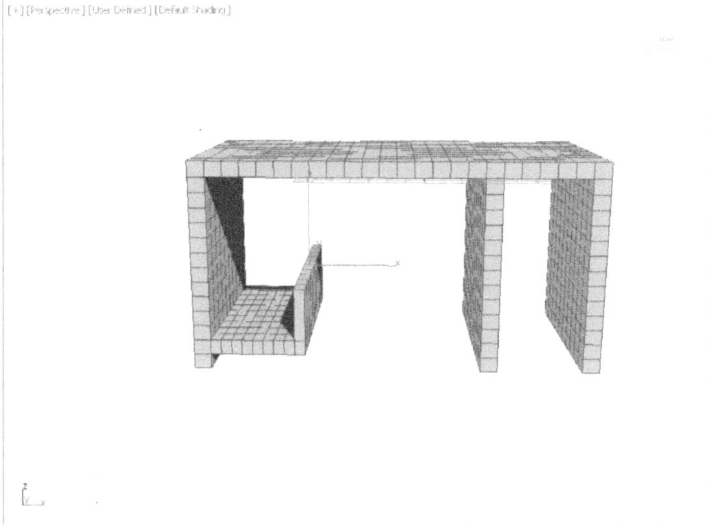

Figure 9-18 *Extruded polygons of computer table*

Creating Top and Bottom Shelves

In this section, you will create top and bottom shelves of *computer table* by using the **Bridge** and **Extrude** tools.

1. Select the polygons from the top of *computer table*, as shown in Figure 9-19.

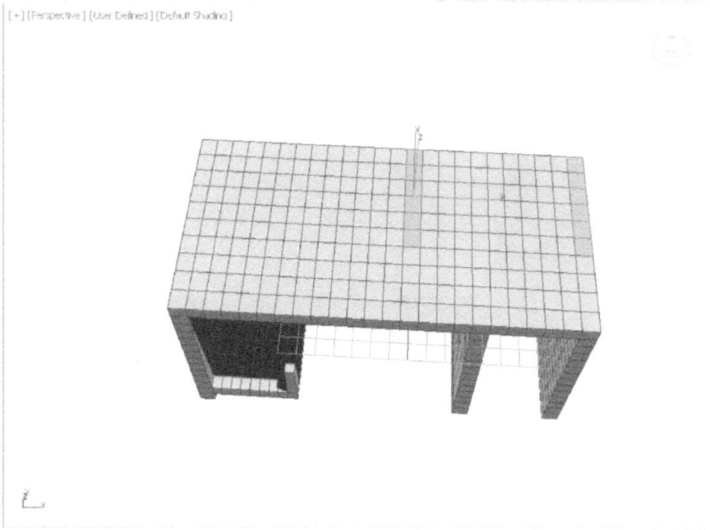

Figure 9-19 *Selected polygons of computer table*

2. Choose the **Extrude Settings** button from the **Extrude** drop-down in the **Polygons** panel; the **Extrude** caddy control is displayed in the viewport. In this caddy control, set the value **60** in the **Height** spinner and choose the **OK** button; the selected polygons are extruded, as shown in Figure 9-20.

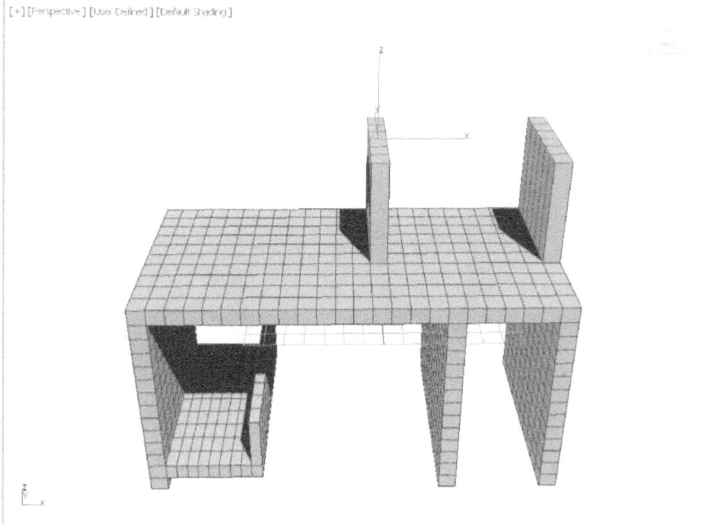

Figure 9-20 *Extruded polygons of computer table*

3. Select the **Edge** sub-object level and then select two edges from the extruded part, as shown in Figure 9-21.

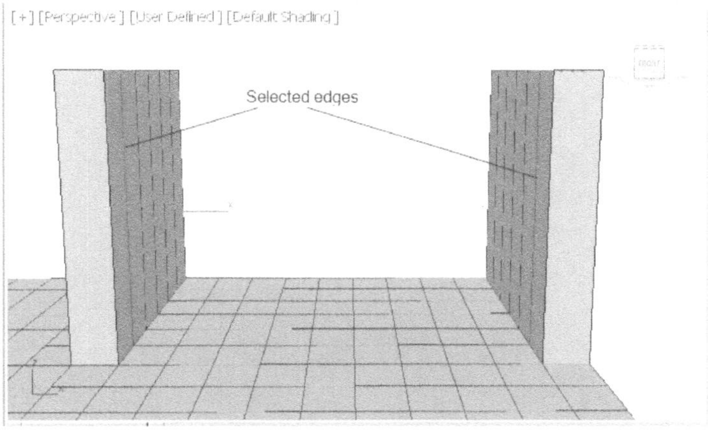

Figure 9-21 *Selected edges of the extruded part*

4. Choose the **Ring** tool from the **Modify Selection** panel; rings of the selected edges are selected.

5. Choose the **Connect Settings** button from the **Connect** drop-down in the **Loops** panel; the **Connect** caddy control is displayed in the viewport.

6. In the **Connect** caddy control, set the value **2** in the **Segments** spinner and **-64** in the **Pinch** spinner, refer to Figure 9-22 and choose the **OK** button; two edges are added at the center of the extruded part, as shown in Figure 9-23.

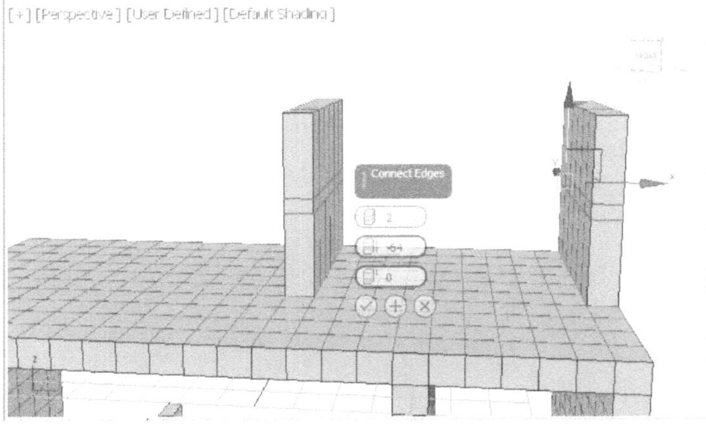

Figure 9-22 *Added edges at the center of the extruded part*

7. Select two edges from the upper part of the extruded portion, as shown in Figure 9-19.

8 Choose the **Insert Loop** tool from the **Loops** panel; two edge loops are added. Choose the **Select and Move** tool and align the edge loops, as shown in Figure 9-24.

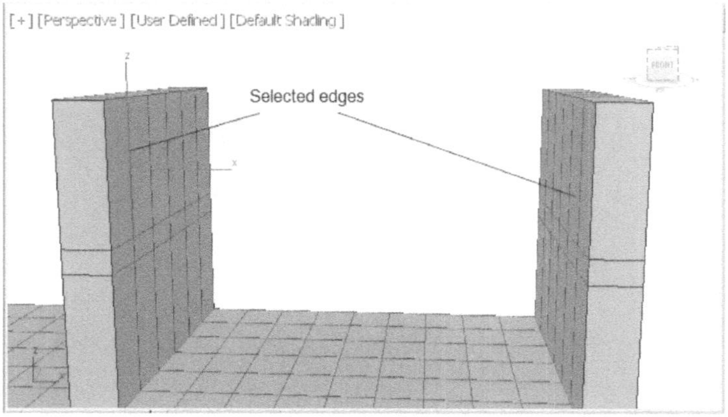

Figure 9-23 *Selected edges from the upper portion of extruded part*

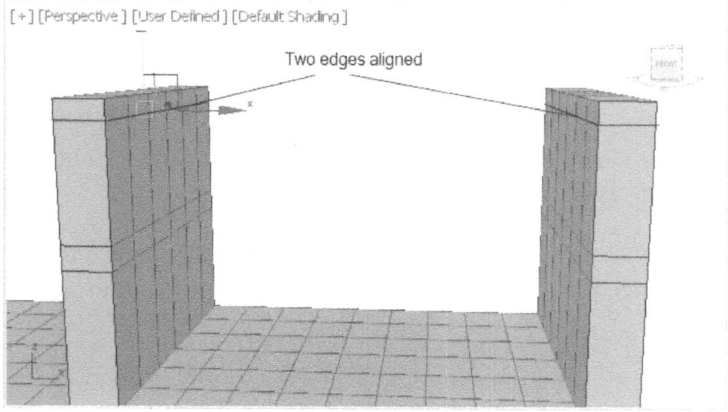

Figure 9-24 *Two edges aligned*

9. Select the **Polygon** sub-object level. Next, select the polygons from the upper extruded part, as shown in Figure 9-25.

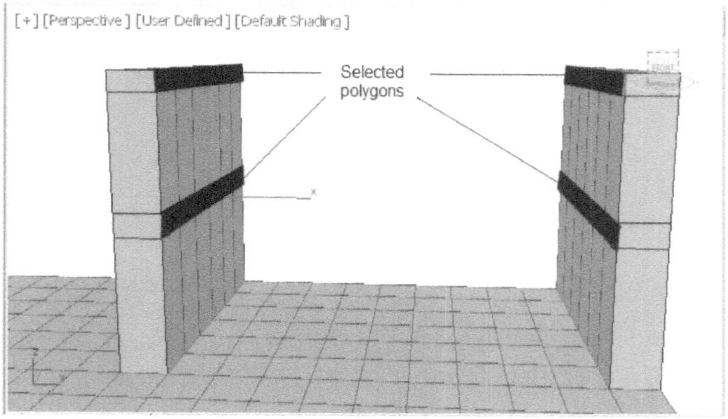

Figure 9-25 *Selected polygons of the extruded part*

10. Choose the **Bridge** tool from the **Polygons** panel; opposite polygons get connected by a bridge, as shown in Figure 9-26.

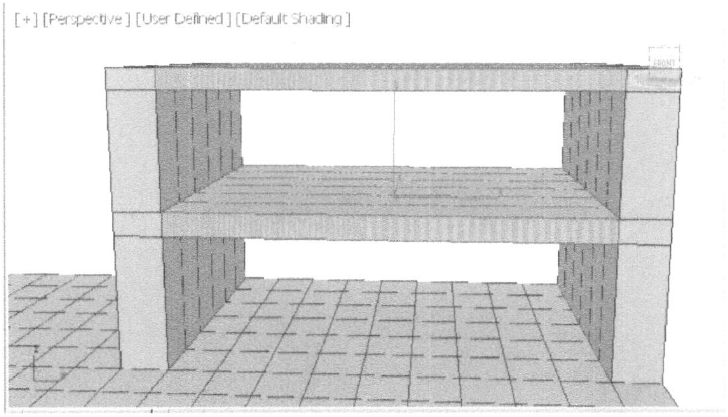

Figure 9-26 *Selected polygons connected by bridges*

Next, you will create lower shelves of *computer table*.

11. Select the polygons from the lower part of computer table, as shown in Figure 9-27.

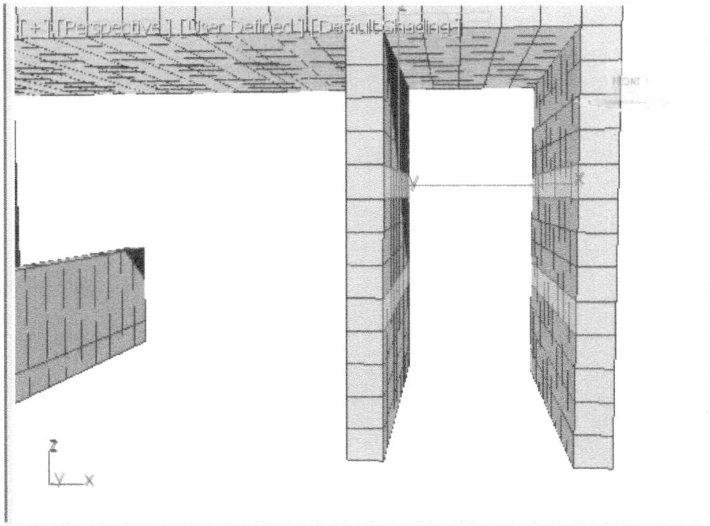

Figure 9-27 *Selected polygons of the lower parts*

12. Choose the **Bridge** tool from the **Polygons** panel; opposite polygons get connected by a bridge, as shown in Figure 9-28.

Creating the Keyboard Support

In this section, you need to create the keyboard support by chamfering edges and extruding polygons.

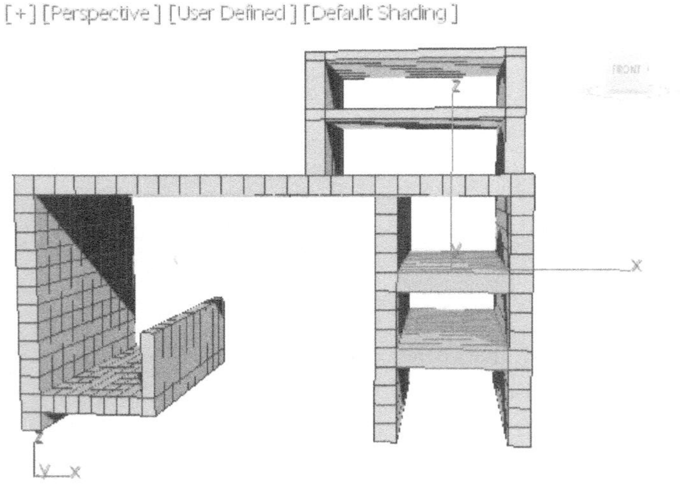

Figure 9-28 *Selected polygons connected by bridge*

1. Select the **Edge** sub-object level and then select two edges from the lower left part of *computer table*, as shown in Figure 9-29.

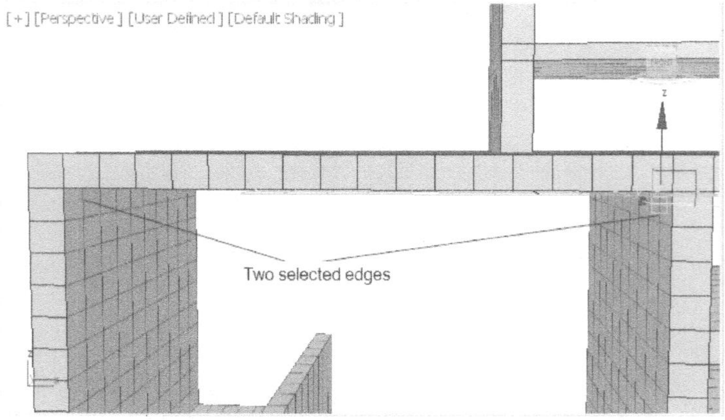

Figure 9-29 *Two edges selected from the lower left part*

2. Choose the **Inset Loop** tool from the **Loops** panel; two edge loops are added, as shown in Figure 9-30.

3. Make sure that the added edge loops are selected. Next, press the CTRL key and select two

more edges from the lower part of *computer table*, as shown in Figure 9-31.

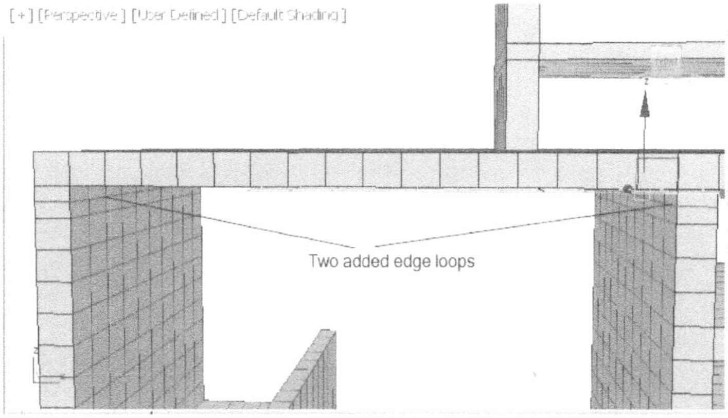

Figure 9-30 *Two added edge loops*

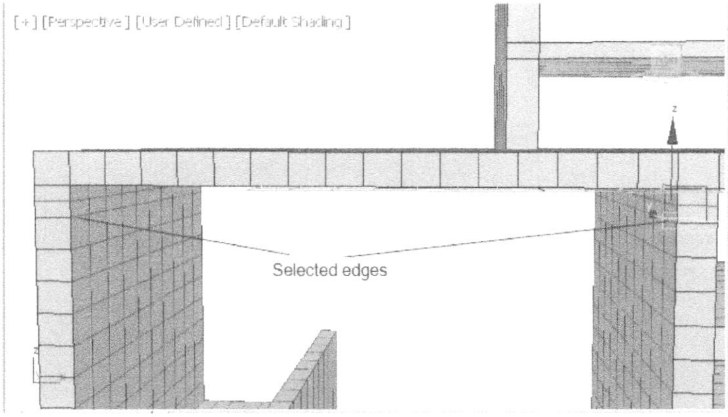

Figure 9-31 *Two edges selected*

4. Choose the **Loop** tool from the **Modify Selection** panel; two edge loops get added to the selection, as shown in Figure 9-32.

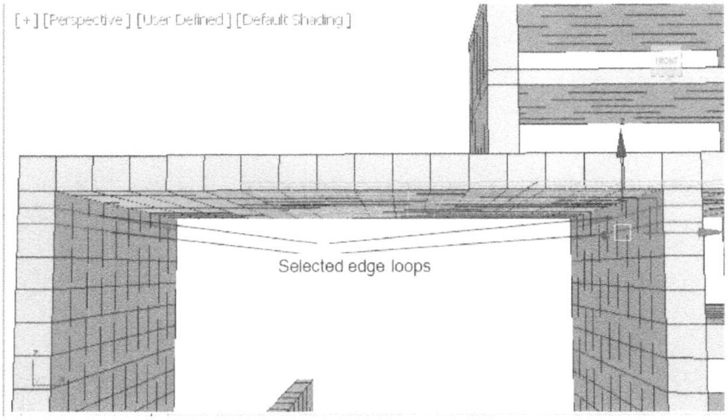

Figure 9-32 *Selected edge loops*

5. Choose the **Chamfer Settings** button from the **Chamfer** drop-down in the **Edges** panel; the **Chamfer** caddy control is displayed in the viewport. In this caddy control, set the value **0.5** in the **Edge Chamfer Amount** spinner and choose the **OK** button; the selected edge loops are chamfered, as shown in Figure 9-33.

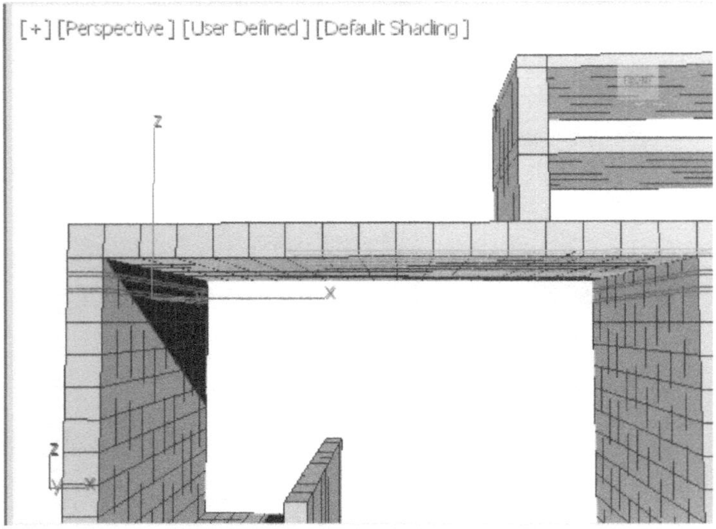

Figure 9-33 *Two edge loops chamfered*

6. Select the **Polygon** sub-object level and then select the polygons between the chamfered edge loops, as shown in Figure 9-34.

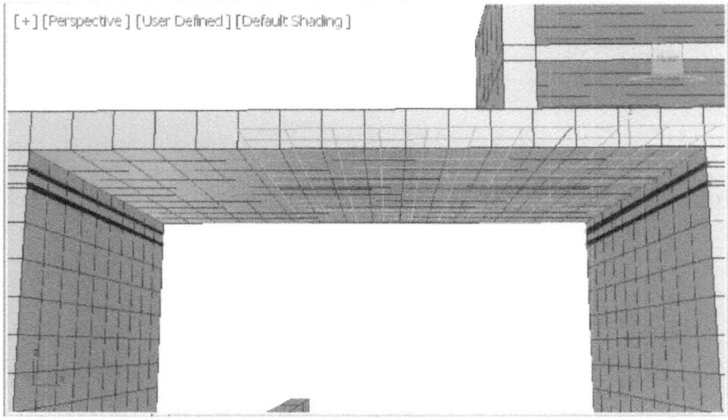

Figure 9-34 *Selected polygons between the chamfered edges*

7. Choose the **Extrude Settings** button from the **Extrude** drop-down in the **Polygons** panel; the **Extrude** caddy control is displayed in the viewport. In this caddy control, set the value **5.5** in the **Height** spinner and choose the **OK** button; the selected polygons are extruded, as shown in Figure 9-35.

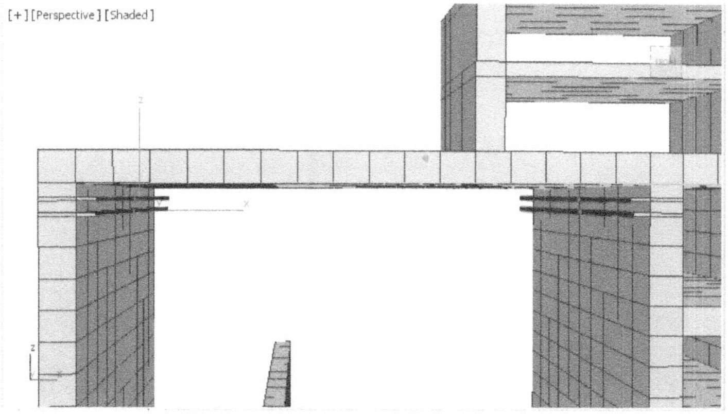

Figure 9-35 *Selected polygons extruded*

8. Activate the Top viewport and choose **Create > geometry** from the **Command Panel**. Now, choose the **Box** tool from the **Object Type** rollout. Next, create a box in the Top viewport.

9. Choose the **Modify** tab from the **Command Panel**. Next, in the **Parameters** rollout, set the parameters as given next:

   Length:**109.45** Width: **160** Height: **2.8**

10. Modify the name of the box as *Keyboard support*. Next, align it in the viewports, as shown in Figure 9-36.

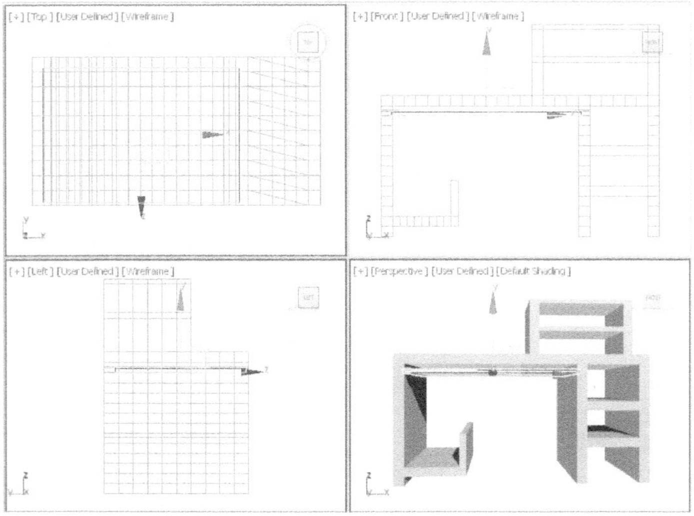

Figure 9-36 *The Keyboard support geometry aligned in the viewports*

11. Now, you need to apply texture to the model. You can download the *c09_3dsmax_2017_tut.zip* file from *www.cadsofttech.com* and extract the contents of the zip file. Next, save *wood.jpg* at the location *\Documents\3dsmax2017\c09_tut1\sceneassets\images* and apply texture to the model, as discussed in the earlier chapters.

Saving and Rendering the Scene

In this section, you will save the scene and then render it. You can view the final rendered image of this model by downloading the *c09_3dsmax_2017_rndr.zip* file from *www.cadsofttech.com*.The path of the file is as follows: *Textbooks > Animation and Visual Effects > 3ds Max > Autodesk 3ds Max 2017 for Novices*.

1. Change the background color of the scene to white, as discussed in Tutorial 1 of Chapter 2. Choose **Save** from the **Application** menu.

2. Activate the Perspective viewport. Next, choose the **Render Production** tool from the **Main Toolbar**; the **Rendered Frame** window is displayed. This window shows the final output of the scene, refer to Figure 9-37.

Figure 9-37 *The rendered model of computer table*

EXERCISE

The rendered output of the scene used in the Exercise 1 can be accessed by downloading the *c09_3dsmax_2017_exr.zip* file from *www.www.cadsofttech.com*.The path of the file is as follows: *Textbooks > Animation and Visual Effects > 3ds Max > Autodesk 3ds Max 2017 for Novices.*

Exercise 1

Create a model of an acoustic guitar and apply texture to it, refer to Figure 9-38.

(Expected time: 50 min)

Figure 9-38 *The model of the guitar*

This page is intentionally left blank

Chapter 10

Compound Objects

Learning Objectives

After completing this chapter, you will be able to:

- *Create compound objects*
- *Modify compound objects*

INTRODUCTION

In this chapter, you will create complex models by combining two or more existing objects into a single object. The different techniques explained in this chapter will help you to combine different 2D and 3D shapes that would otherwise be difficult or impossible to do.

COMPOUND OBJECTS

In 3ds Max, the compound objects are created by combining two or more objects. There are twelve tools to create various compound objects. To create these compound objects, you need to choose the respective tools. To create a compound object, choose **Create > Geometry** in the **Command Panel**; a drop-down list will be displayed below the **Geometry** button. Select the **Compound Objects** option from the drop-down list; the **Object Type** rollout will be displayed with all the tools to create compound objects, refer to Figure 10-1. The most commonly used tools to create compound objects are discussed next.

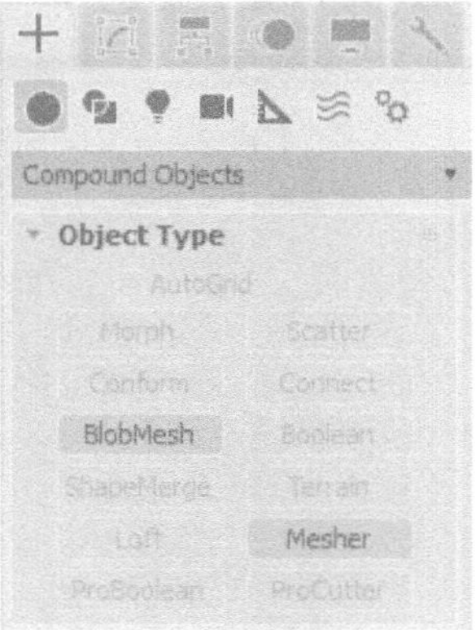

***Figure 10-1** The tools for creating compound objects displayed in the rollout*

Boolean

Menu bar: Create > Compound > Boolean
Toolbar: Create > Geometry > Compound Objects > Object Type rollout > Boolean

The **Boolean** tool is used to trim or combine objects in an operation. To carry out boolean operation, you need two objects. The objects to be operated upon are known as operands. To create a boolean object, first create two overlapping objects in the viewport, refer to Figure 10-2. Select one of the objects in the viewport. Choose **Create > Geometry** in the **Command Panel**; a drop-down list will be displayed. Select the **Compound Objects** option from the drop-down list and choose the **Boolean** tool from the **Object Type** rollout; the **Name and Color**, **Boolean Parameters**, and **Operand Parameters** rollouts will be displayed, refer to Figure 10-3.

Terrain

Menu bar: Create > Compound > Terrain
Toolbar: Create > Geometry > Compound Objects > Object Type rollout > Terrain

The terrain compound objects are created from splines. In this case, the splines represent the contour lines. The contour lines are the lines drawn on a map, which connect the points of equal height. Usually, the terrain objects are used to add buildings and roads to them to create virtual cities. To create a terrain object, first create the contour lines in the Top viewport using the **Line** tool, as shown in Figure 10-3. Align these lines in the Front viewport, as shown in Figure 10-4.

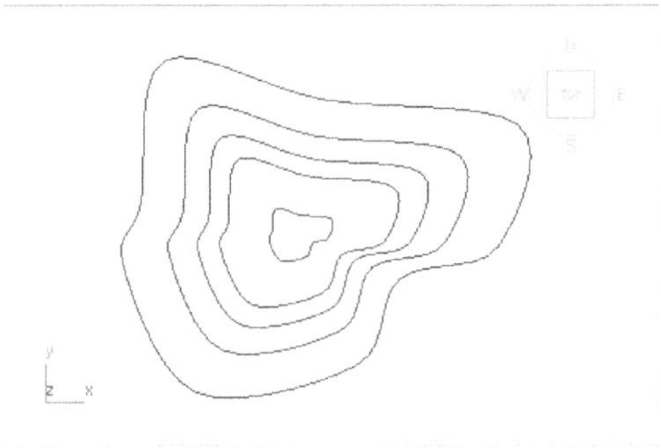

Figure 10-2 The contour lines created for the terrain object in the Top viewport

Figure 10-3 The contour lines aligned for the terrain object in the Front viewport

You can also import an AutoCAD drawing file to use as contour data. Next, select all the contour lines and choose the **Terrain** tool from the **Object Type** rollout of the **Compound Objects**; a 3D terrain object will be created. The **Name and Color**, **Pick Operand**, **Parameters**, **Simplification**, and **Color by Elevation** rollouts will also be displayed to modify the terrain object.

Scatter

Menu bar: Create > Compound > Scatter
Toolbar: Create > Geometry > Compound Objects > Object Type rollout > Scatter

The **Scatter** tool is used to randomly distribute the source object as an array or over the surface of the distribution object. To create a scatter object, first create an object to be used as the source object and create another object to be used as the distribution object. Next, select the source object and choose the **Scatter** tool from the **Object Type** rollout of the **Compound Objects**; the **Pick Distribution Object**, **Scatter Objects**, **Transforms**, **Display**, and **Load/Save Presets** rollouts will be displayed. Now, choose the **Pick Distribution Object** button in the **Pick Distribution Object** rollout and move the cursor over the distribution object; the shape of the cursor will change into a selection cursor. Select the distribution object in the viewport. Next, expand the **Scatter Objects** rollout. In the **Source Object Parameters** area, set the required value in the **Duplicates** spinner to create a number of duplicates of the source object, refer to Figures 10-4 and 10-5.

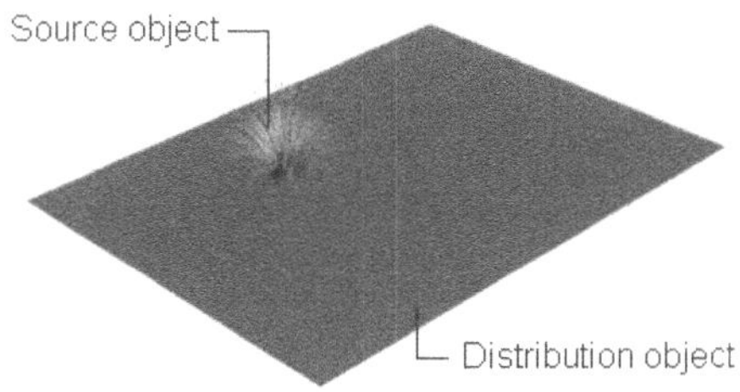

Figure 10-4 The distribution and source objects

Figure 10-5 The duplicates of the source object distributed over the distribution object

Connect

Menu bar:	Create > Compound > Connect
Toolbar:	Create > Geometry > Compound Objects > Object Type rollout > Connect

The **Connect** tool is used to connect two or more objects at the holes on their surfaces. To connect the objects, create two objects and then convert them into editable objects. Next, select the **Polygon** or **Face** sub-object level in the modifier stack. Now, select the sub-objects of the editable objects in the viewport and then press the DELETE key to delete the selected face or polygon sub-objects, refer to Figure 10-6. Note that you need to position the objects in such a way that the holes should face each other, as shown in Figure 10-6. Next, select one of the objects and choose the **Connect** tool from the **Object Type** rollout in the **Compound Objects**. Choose the **Pick Operand** button in the **Pick Operand** rollout and move the cursor over the second object in the viewport; the shape of the cursor will change into a selection cursor. Select another object; new faces will be created to connect the holes of the objects, as shown in Figure 10-7. Next, right-click to exit the command.

When you choose the **Connect** tool, the **Name and Color**, **Pick Operand**, **Parameters**, and **Display/Update** rollouts will be displayed, refer to Figure 10-8.

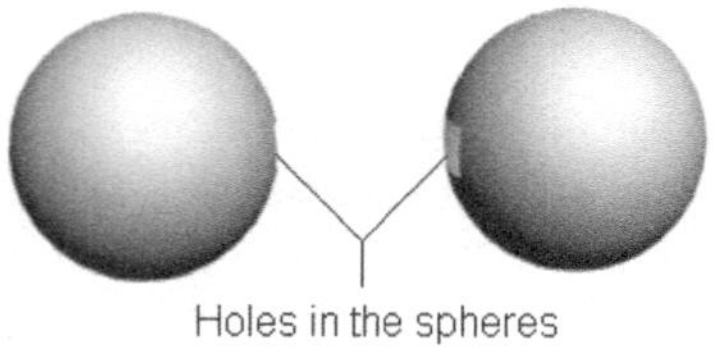

Figure10-6 *Two spheres with holes on their surfaces*

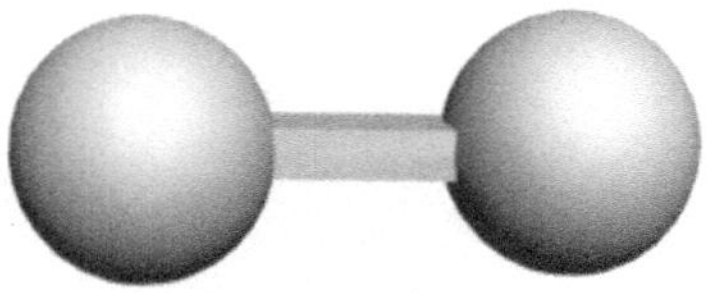

Figure 10-7 *The spheres after performing the **Connect** operation*

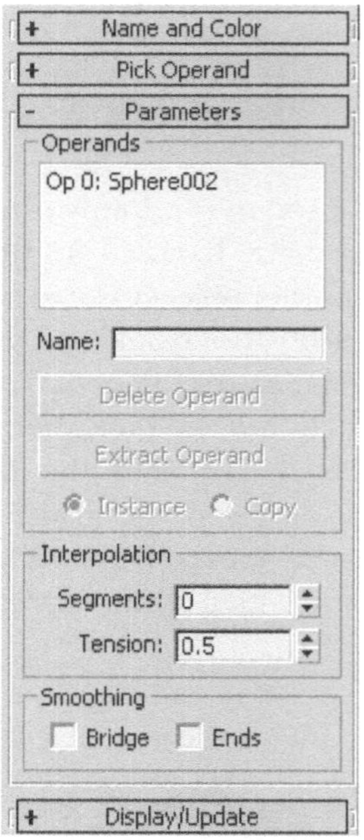

Figure 10-8 *The rollouts displayed after choosing the **Connect** tool*

Loft

Menu bar:	Create > Compound > Loft
Toolbar:	Create > Geometry > Compound Objects > Object Type rollout > Loft

The **Loft** tool is used to create 3D geometry or loft objects from 2D shapes. For creating a loft object, you need to create a path and shape spline in the viewport.

To create a loft object, choose **Create > Shapes > Splines** in the **Command Panel**. Next, create a spline that you need to loft and create another spline to be used as a path, refer to Figure 10-9. Select the spline in the viewport that you want to loft along the path. Choose the **Loft** tool from the **Object Type** rollout in the **Compound Objects**; the **Name and Color**, **Creation Method**, **Surface Parameters**, **Path Parameters**, and **Skin Parameters** rollouts will be displayed, as shown in Figure 10-10. Now, choose the **Get Path** button in the **Creation Method** rollout and move the cursor over the path spline in the viewport; the shape of the cursor will change into a selection cursor. Select the path spline; the loft object will be created along the shape of the path spline, refer to Figure 10-11.

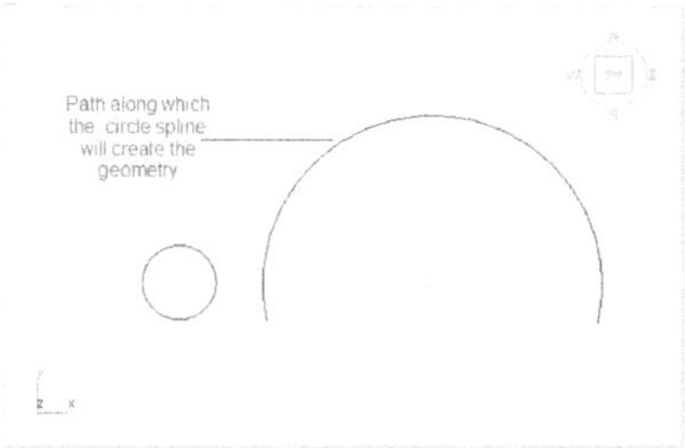

Figure 10-9 Two splines to create a loft object

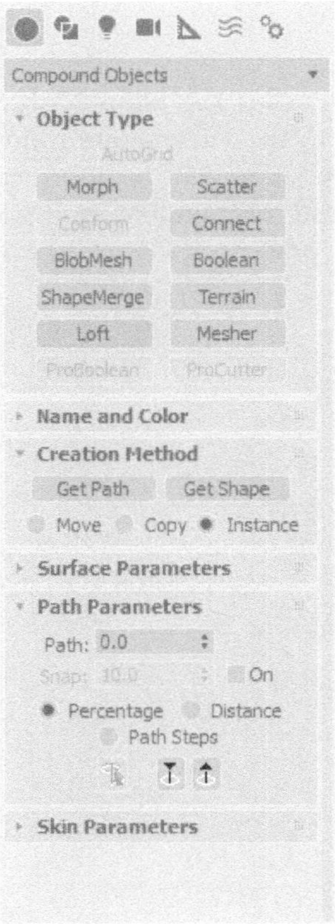

*Figure 10-10 Partial view of the rollouts displayed after invoking the **Loft** tool*

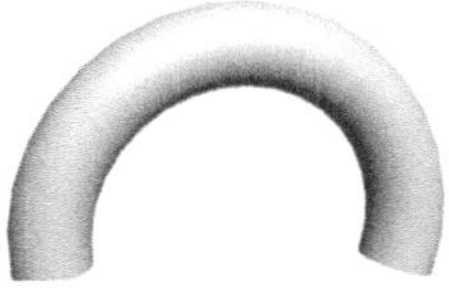

Figure 10-11 The loft object created

You can also create a loft object using the **Get Shape** button in the **Creation Method** rollout. To do so, first you need to select the path spline, and then choose the **Loft** tool from the **Object Type** rollout in the **Compound Objects**. Choose the **Get Shape** button in the **Creation Method** rollout; a selection cursor will be displayed in the viewport, which will prompt you to select the shape spline. Select the shape spline; the loft object will be created along the path.

ShapeMerge

Menu bar:	Create > Compound > ShapeMerge
Toolbar:	Create > Geometry > Compound Objects > Object Type rollout > ShapeMerge

In a shape merge object, the shape is either embedded in the mesh object or subtracted from it. To create a shape merge object, first you need to create a mesh object and a shape, and then align them, as shown in Figure 10-12. Next, select the mesh object and choose the **ShapeMerge** tool from the **Object Type** rollout of the **Compound Objects**. Next, choose the **Pick Shape** button in the **Pick Operand** rollout; a selection cursor will be displayed that will prompt you to select the shape spline. Select the shape spline and right-click to exit the shape merge command.

To view the shape merge object with the embedded shape in it, select the shape merge object. Choose the **Modify** tab in the **Command Panel** and select the **Face Extrude** modifier from the **Modifier List** drop-down list. In the **Parameters** rollout of the **Face Extrude** modifier, set the value in the **Amount** spinner; the shape will be embedded in the mesh object, as shown in Figure 10-13.

To view the subtracted shape in the shape merge object, select the **ShapeMerge** object level in the modifier stack; various rollouts will be displayed. In the **Parameters** rollout, select the **Cookie Cutter** radio button in the **Operation** area; the shape will be subtracted from the mesh object, as shown in Figure 10-14. When you choose the **ShapeMerge** tool, the **Name and Color**, **Pick Operand**, **Parameters**, and **Display/Update** rollouts will be displayed, as shown in Figure 10-15.

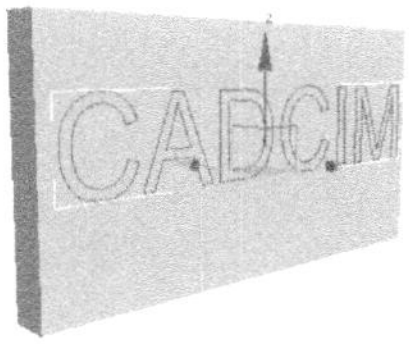

***Figure 10-12** Alignment of the shape with the mesh object*

***Figure 10-13** The shape embedded in the mesh object*

***Figure 10-14** The shape subtracted from the mesh object*

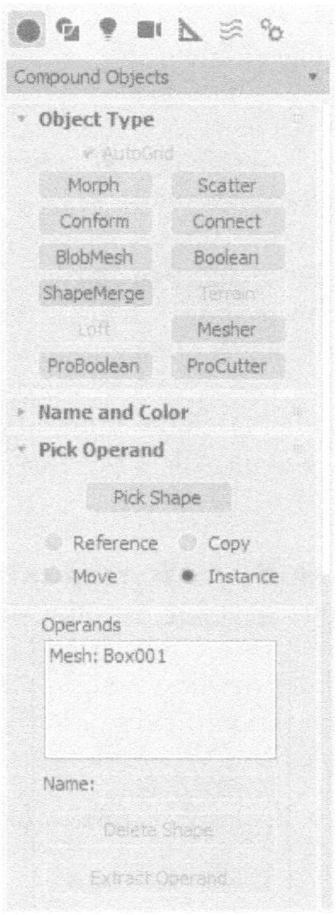

***Figure 10-15** Partial view of the rollouts displayed after invoking the **ShapeMerge** tool*

TUTORIALS

Before starting the tutorials, you need to download the *c10_3dsmax_2017_tut.zip* file from *www.cadsofttech.com*. Extract the contents of the zipped file and save them in the *Documents* folder.

Tutorial 1

In this tutorial, you will create a toothpaste tube using the **Loft** and **Boolean** tools, as shown in Figure 10-16. **(Expected time: 30 min)**

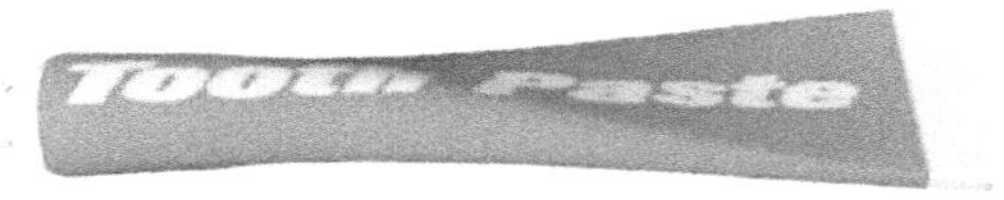

***Figure 10-16** The model of a toothpaste tube*

The following steps are required to complete this tutorial:

a. Create the project folder.
b. Create the body of toothpaste tube.
c. Create the cap of toothpaste tube.
d. Assign materials and maps to the tube.
e. Save and render the scene.

Creating the Project Folder

1. Create a new project folder with the name *c10_tut1* at *\Documents\3dsmax2017* and then save the file with the name *c10tut1*, as discussed in Tutorial 1 of Chapter 2.

2. Open the Windows Explorer and then browse to the *c10_3dsmax_2017_tut* folder and copy the *toothpaste_tube.jpg* file from this folder to *\Documents\3dsmax2017\c10_tut1\sceneassets\images*.

Creating the Body of Toothpaste Tube

In this section, you will create the body of toothpaste tube by using the **Loft** tool.

1. Activate the Front viewport. Choose **Create > Shapes** from the **Command Panel**; the **Splines** option is displayed in the drop-down list in the **Command Panel**. Next, choose the **Circle** tool from the **Object Type** rollout.

2. Create a circle in the viewport to use it as a shape spline. It is automatically named as *Circle001*.

3. In the **Parameters** rollout, set the value **4.0** in the **Radius** spinner.

4. Create another circle in the Front viewport, refer to Figure 10-60, to use as a shape spline and set the value **16.0** in the **Radius** spinner. It is automatically named as *Circle002*.

5. Choose the **Line** tool from **Create > Shapes > Splines > Object Type** rollout in the **Command Panel**.

6. Create a line in the Top viewport to use as a path spline. You can assume your own dimensions for the length of the line, refer to Figure 10-17. The line is automatically named as *Line001*.

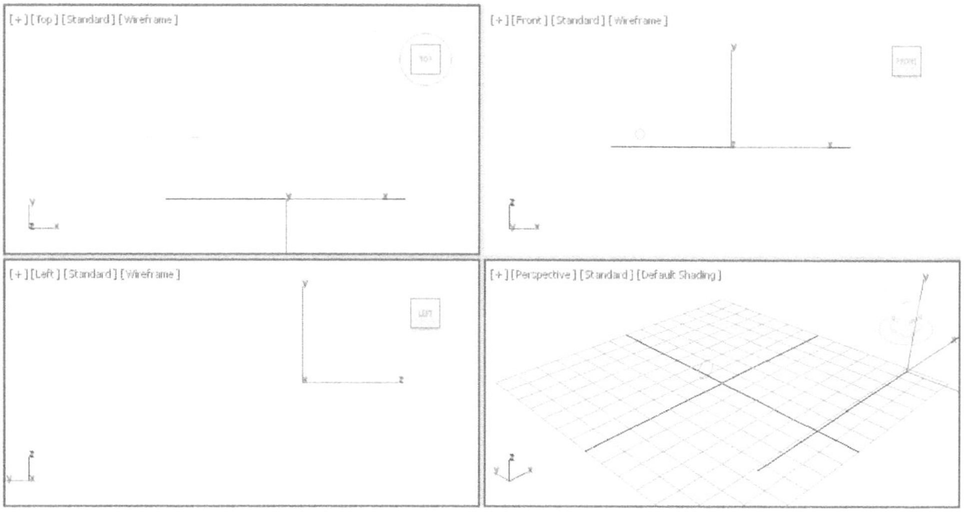

Figure 10-17 *The Circle001, Circle002, and Line001 splines in viewports*

7. Activate the Front viewport and choose the **Rectangle** tool from **Create > Shapes > Splines > Object Type** rollout in the **Command Panel**. Create a rectangle to use as a shape spline. The rectangle is automatically named as *Rectangle001*.

8. In the **Parameters** rollout of *Rectangle001*, set the following parameters:

   Width: **55.033** Length: **5.76** Corner Radius: **1.0**

   The *Rectangle001* spline is displayed in the viewports, as shown in Figure 10-18.

   Next, you will create a loft object using the shape splines, path spline, and the **Loft** tool. You will use *Circle001*, *Circle002*, and *Rectangle001* as the shape splines and *Line001* as the path spline.

9. Activate the Front viewport, select *Line001* spline in the viewport.

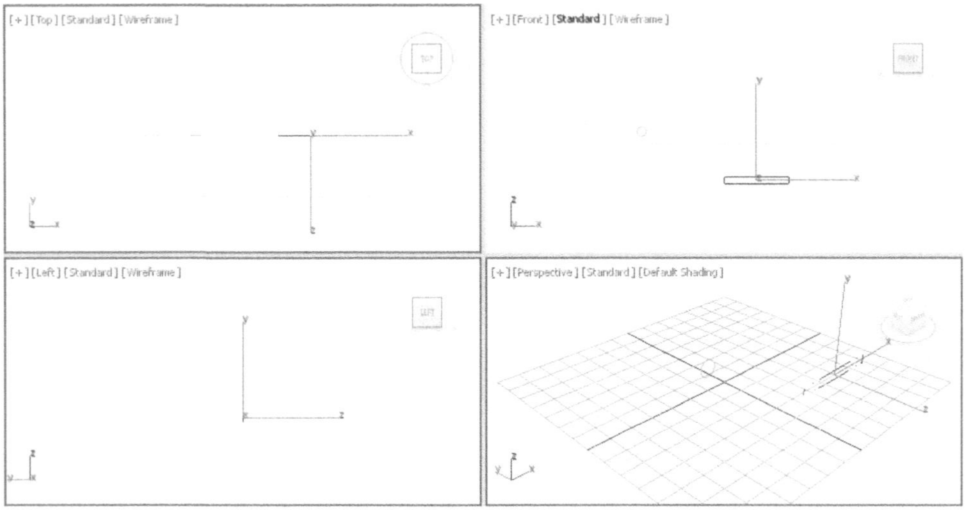

Figure 10-18 *The Rectangle001 displayed in viewports*

10. Choose **Create > Geometry** from the **Command Panel**. Select **Compound Objects** from the drop-down list and then choose the **Loft** tool from the **Object Type** rollout; various rollouts are displayed in the **Command Panel**.

11. In the **Creation Method** rollout, choose the **Get Shape** button and move the cursor over *Circle001* in the Front viewport; the shape of the cursor changes. Now, select *Circle001*; a shape is created, as shown in Figure 10-19. The new shape is automatically named as *Loft001*.

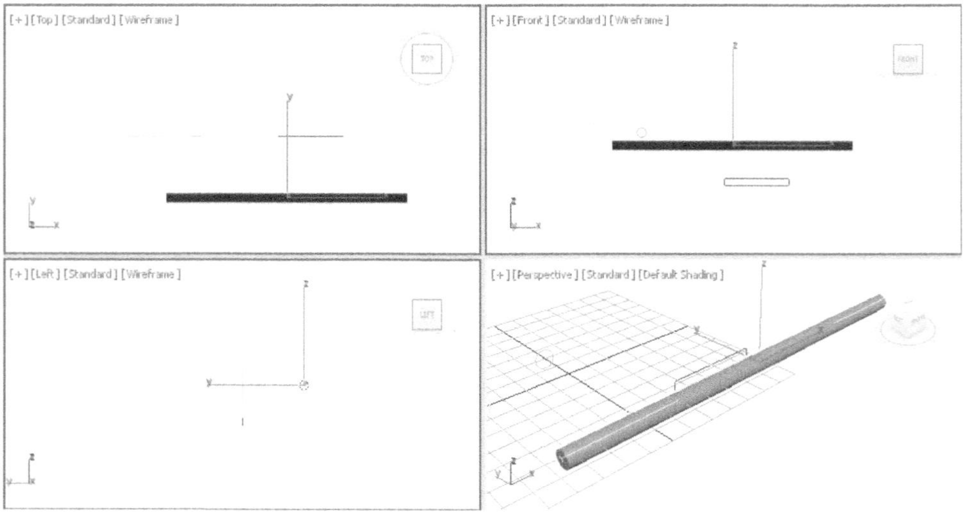

Figure 10-19 *A shape created along the path spline (Line001)*

12. Make sure that *Loft001* is selected in the viewport. In the **Path Parameters** rollout, set the value **8.0** in the **Path** spinner to define the distance between *Circle001* and *Circle002* along *Line001*.

Note

*The value in the **Path** spinner is set according to the length of the line spline.*

13. In the **Creation Method** rollout, choose the **Get Shape** button and move the cursor over *Circle002* in the Front viewport; the shape of the cursor changes. Select *Circle002*; the shape of the tube is created, refer to Figure 10-20. Again, choose the **Get Shape** button to exit the command.

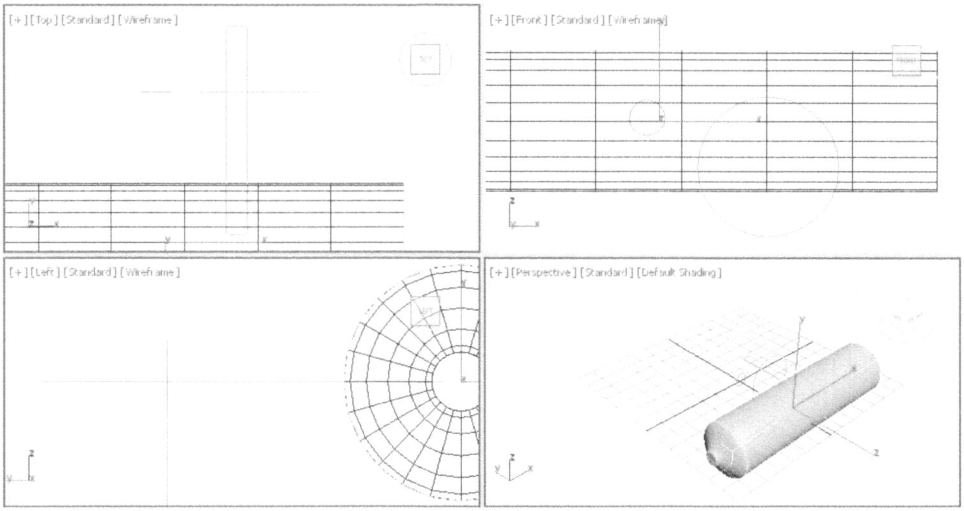

Figure 10-20 *The shape of the tube created along the path spline*

Next, you need to create the shape of the lower portion of the tube. To do so, activate the Top viewport.

14. Make sure that *Loft001* is selected in the viewport. In the **Path Parameters** rollout, set the value **98.0** in the **Path** spinner.

15. In the **Creation Method** rollout, choose the **Get Shape** button and move the cursor over *Rectangle001* in the Front viewport; the shape of the cursor changes. Select *Rectangle001*; the shape of tube is created, refer to Figure 10-21. Next, right-click in the viewport to exit the loft command.

16. Make sure that *Loft001* is selected. In the **Name and Color** rollout, rename *Loft001* to *toothpaste tube*.

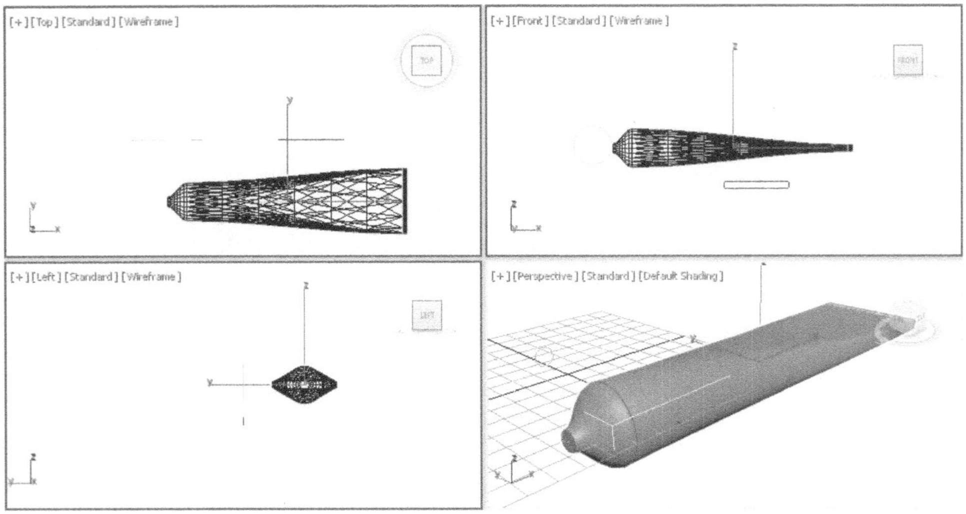

Figure 10-21 The shape of tube created in viewports

Creating the Cap of the Toothpaste Tube

In this section, you will create the cap of the toothpaste tube by using the **Cone** tool.

1. Activate the Left viewport, choose **Create > Geometry** in the **Command Panel** and select the **Standard Primitives** option from the drop-down list. Now, choose the **Cone** tool from the **Object Type** rollout and create a cone.

2. Choose the **Modify** tab from the **Command Panel**. Next, in the **Parameters** rollout, set the following parameters:

   Radius 1: **9.0** Radius 2: **4.0** Height: **20**

   Use the default values for other options.

3. Modify the name of cone as *cap01*.

4. Make sure that *cap01* is selected. Next, right-click on it; a quad menu is displayed. Choose the **Hide Unselected** option to temporarily hide other objects in the viewport.

5. Choose the **Zoom Extents All Selected** tool; *cap01* is zoomed in the viewports, as shown in Figure 10-22.

   Next, you need to create a hole in *cap01* to create the space for inserting the top portion of *toothpaste tube*. To do so, you need to use the **Boolean** tool in **Compound Objects**.

6. Activate the Top viewport and select *cap01*.

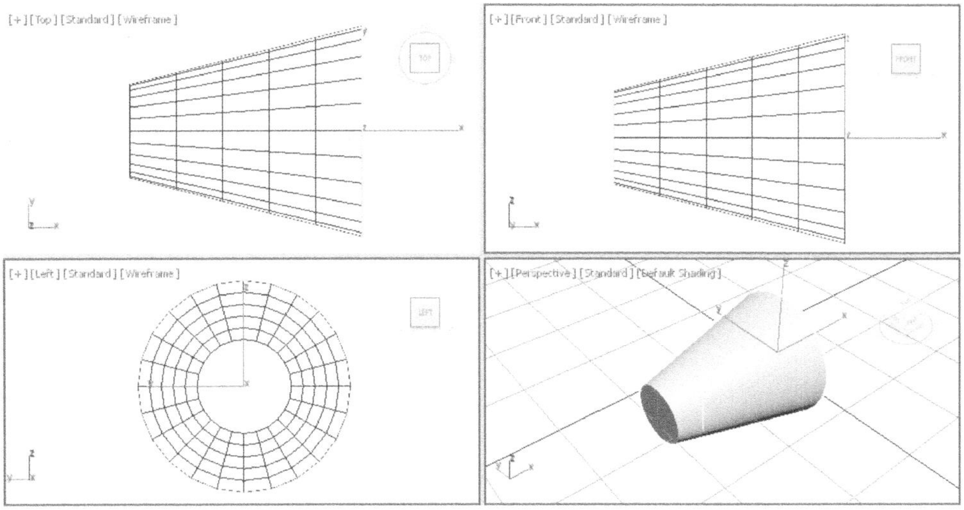

Figure 10-22 The cap01 geometry zoomed in viewports

7. Choose the **Use Selection Center** tool from the Use Pivot Center flyout in the **Main Toolbar**; the coordinate gizmo moves to the center of *cap01*. Also, choose the **Select and Uniform Scale** tool.

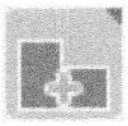

Note

The tools available in the ***Use Center*** *flyout determine the geometric center for the scale and rotate operations.*

8. Activate the Perspective viewport and make sure the cap is selected. Press and hold the SHIFT key and the left mouse button and move the cursor down until the value in the X, Y, and Z coordinates in the coordinate display becomes **90** and then release the left mouse button and the SHIFT key; the **Clone Options** dialog box is displayed. Make sure that the **Copy** radio button is selected and **1** is set in the **Number of Copies** spinner and then choose the **OK** button; *cap002* is displayed, as shown in Figure 10-23. You need to switch to wireframe mode in the Perspective viewport to view *cap002*, refer to Figure 10-23.

9. Activate the Front viewport and then choose the **Select and Move** tool and align *cap002* with *cap01*, as shown in Figure 10-24.

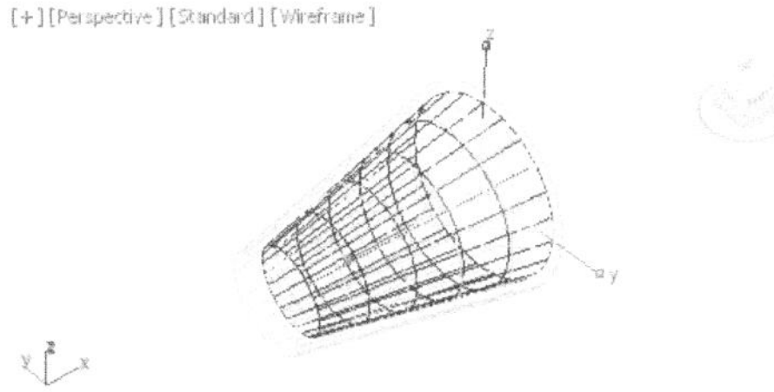

Figure 10-23 The cap002 geometry displayed

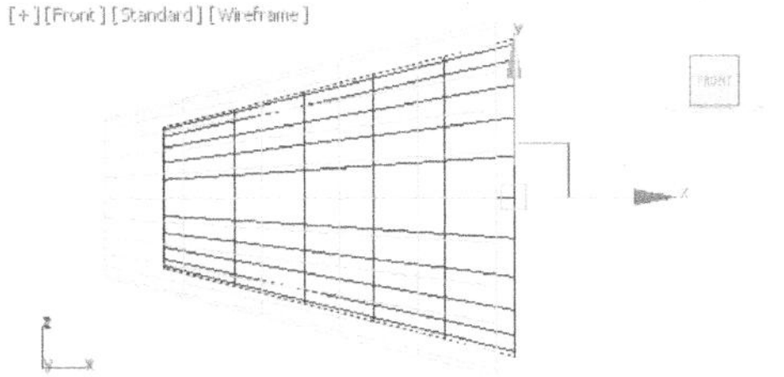

Figure 10-24 The cap002 geometry after the alignment

10. Select *cap01* in the Top viewport. Choose **Create > Geometry** in the **Command Panel**. Select the **Compound Objects** option from the drop-down list in the **Command Panel** and choose the **Boolean** tool from the **Object Type** rollout.

11. Choose the **Add Operands** button from the **Boolean Parameters** rollout and then click on *cap002* in a viewport. Choose the **Subtract** button from the **Operand Parameters** rollout; the overlapping portion of *cap002* is removed from *cap01*. Next, switch to Shaded mode in the Perspective viewport. Figure 10-25 shows *cap01* in the viewports. Next, right-click to exit the boolean command.

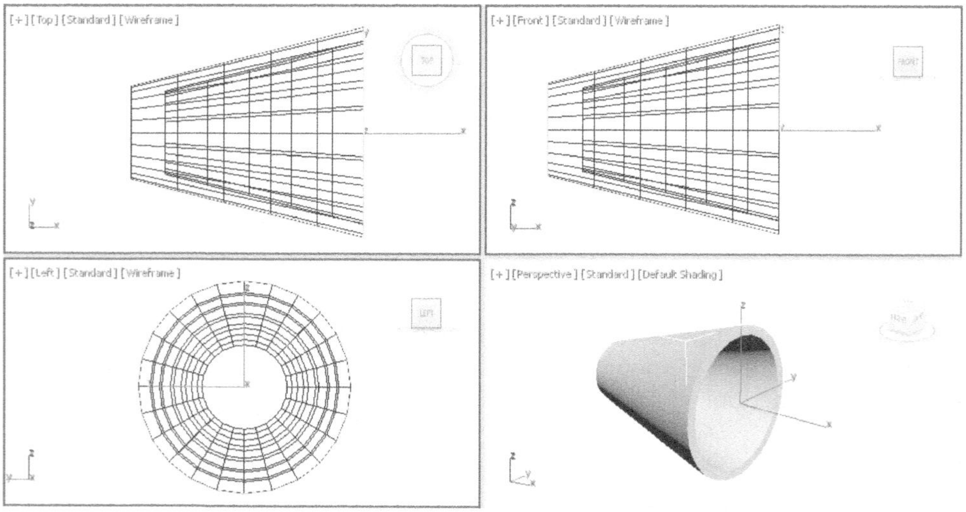

Figure 10-25 *The cap01 after performing the Subtraction(A-B) boolean operation in viewports*

12. In the **Pick Boolean** rollout, select the **Move** radio button, if it is not already selected. Choose the **Pick Operand B** button and move the cursor over *cap002*; the shape of the cursor changes. Select *cap002*; the overlapping portion of *cap002* is removed from *cap01*.

Note

To view the effect of the boolean operation in the Perspective viewport, you need to use the ***Orbit*** *tool.*

Next, you need to modify the cap using its sub-objects to make it look realistic.

13. Convert *cap01* into editable mesh object.

14. In the modifier stack, click on the plus (+) sign on the left side of **Editable Mesh**; all the sub-object levels are displayed.

15. Select the **Polygon** sub-object level to activate it.

16. In the **Selection** rollout, select the **Ignore Backfacing** check box.

17. In the Left viewport, select the alternative row of polygons of *cap01* by holding the CTRL key, as shown in Figure 10-26.

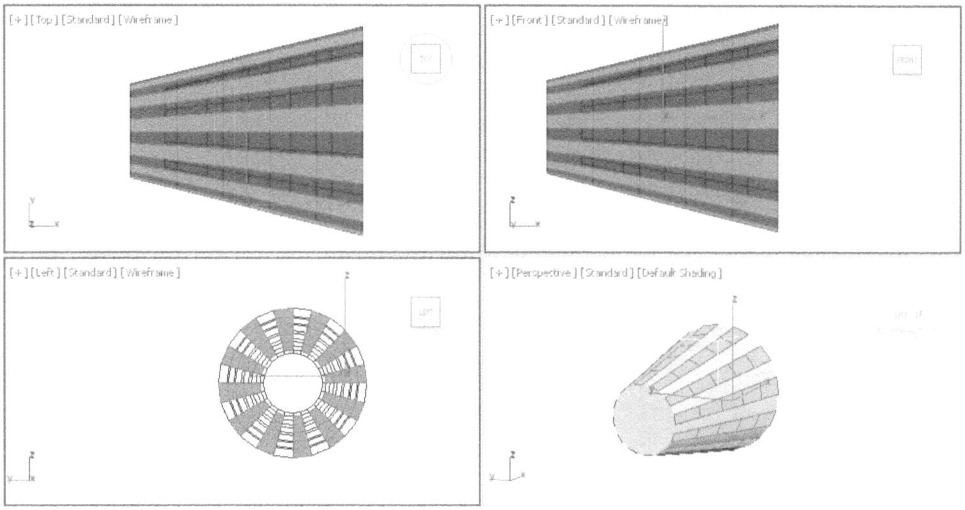

Figure 10-26 *The selected polygons of cap01*

18. Create a selection set of the selected polygons.

Note
*You can use the selection set of the selected objects or sub-objects, if needed. To create a selection set, enter **selected polygons** in the **Named Selection Sets** drop-down list in the **Main Toolbar** and press ENTER, refer to Figure 10-27.*

19. In the **Edit Geometry** rollout, set the value **0.1** in the **Extrude** spinner, and then press the ENTER key; the selected polygons are extruded.

20. The extruded polygons are displayed, as shown in Figure 10-28. In the modifier stack, select the **Polygon** sub-object level again to deactivate it.

Figure 10-27 *The **Named Selection Sets** drop-down list*

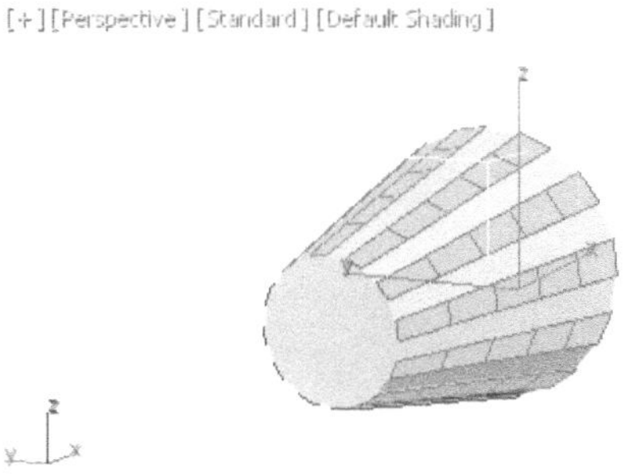

Figure 10-28 *The extruded polygons in the Perspective viewport*

21. Right-click in any of the viewports; a quad menu is displayed. Choose the **Unhide All** option; all the objects and shapes are displayed in the viewport.

22. Choose the **Zoom Extents All** tool; all the objects are displayed, as shown in Figure 10-29.

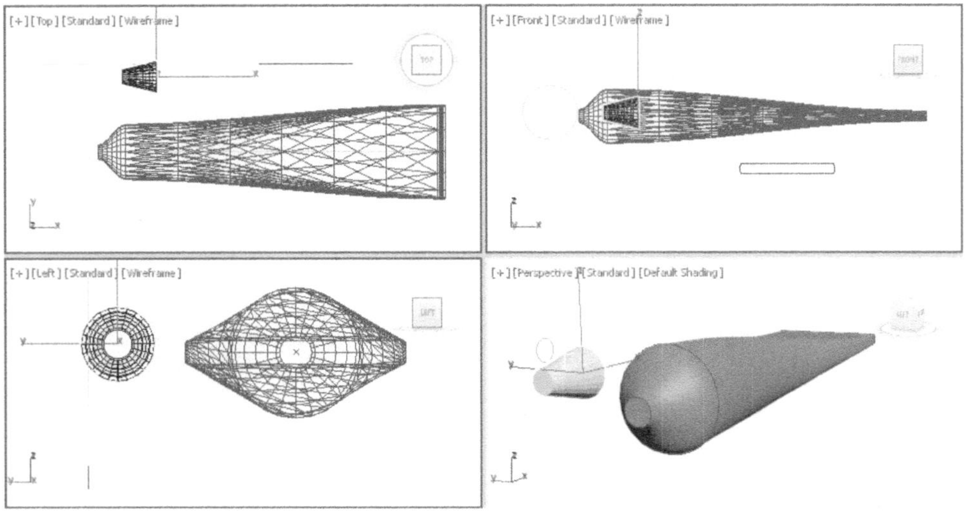

Figure 10-29 *The objects in the viewports*

23. Make sure *cap01* is selected and then choose the **Use Pivot Point Center** tool from the **Main Toolbar**. Now, align *cap01* with *toothpaste tube* using the **Select and Move** tool in viewports, as shown in Figure 10-30.

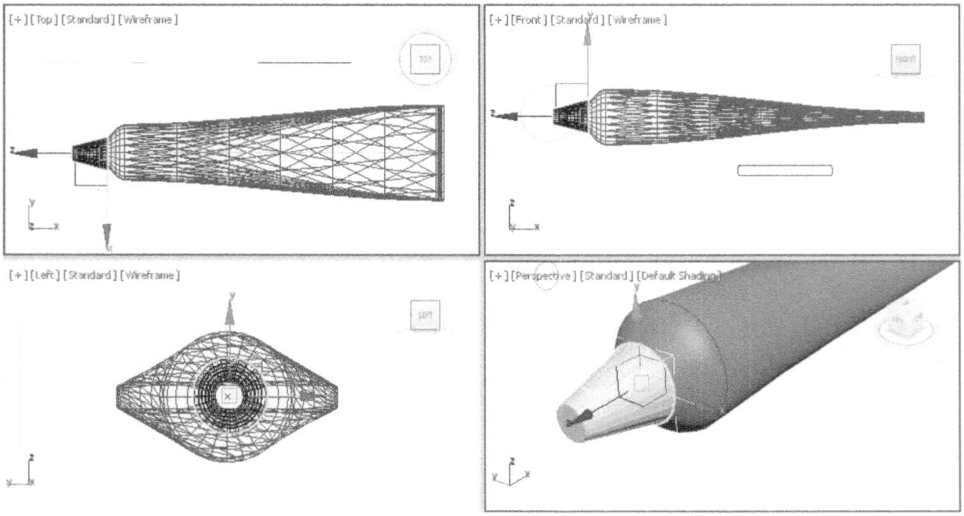

Figure 10-30 *The cap01 geometry in the viewports after being aligned with toothpaste tube*

24. Select *Circle001*, *Circle002*, *Line001*, and *Rectangle001* shapes from the Scene Explorer using the CTRL key. Next, press the DELETE key to delete all the shapes from the scene.

Assigning Materials and Maps to the Tube

In this section, you will assign the materials and maps to *toothpaste tube*.

1. Select *toothpaste tube* in any of the viewports. Next, choose **Rendering > Material Editor > Compact Material Editor** from the menu bar; the **Material Editor** dialog box is displayed. Make sure the first sample slot is selected in the **Material Editor** dialog box. By default, the **Standard** material is selected in it and you need to use the same for *toothpaste tube*.

2. Make sure that the **Blinn** shader is selected in the **Shader Basic Parameters** rollout. In the **Blinn Basic Parameters** rollout, choose the **Diffuse** map button on the right side of the **Diffuse** color swatch; the **Material/Map Browser** dialog box is displayed. Expand the **Maps > General** rollout and select the **Bitmap** map. Next, choose the **OK** button; the **Select Bitmap Image File** dialog box is displayed. As the project folder is already set, the *images* folder is displayed in the **Look in** drop-down list of this dialog box.

3. Select the file *toothpaste_tube.jpg* and choose the **Open** button; the image is displayed in the selected sample slot. Also, the rollouts to modify the bitmap are displayed in the **Material Editor** dialog box. Now, choose the **Go to Parent** tool to go back to the parent level.

4. In the **Specular Highlights** area, set the following parameters:

   Specular Level: **56** Glossiness: **40**

5. Make sure that *toothpaste tube* is selected in the viewport. Next, choose the **Assign Material to Selection** tool in the **Material Editor** dialog box. Also, choose the **Show Shaded Material in Viewport** tool to display the map in the viewport.

6. You will notice that the map in *toothpaste tube* is not showing properly, refer to Figure 10-31. To align the map properly, you need to apply the **UVW Map** modifier.

7. Choose the **Modify** tab in the **Command Panel**. Select the **UVW Map** modifier from the **Modifier List** drop-down list; the **UVW Map** modifier is displayed in the modifier stack. Also, the **Parameters** rollout is displayed in the Modify panel.

8. In the **Parameters** rollout, select the **Box** radio button.

9. In the **Alignment** area of the **Parameters** rollout, select the **Z** radio button and set the value **0.5** in the **Height** spinner. You can set values in the **Length** and **Width** spinners and align the gizmo as per the length and width of the tube to align the map properly, refer to Figure 10-32.

   Next, you will create the material for *cap01*.

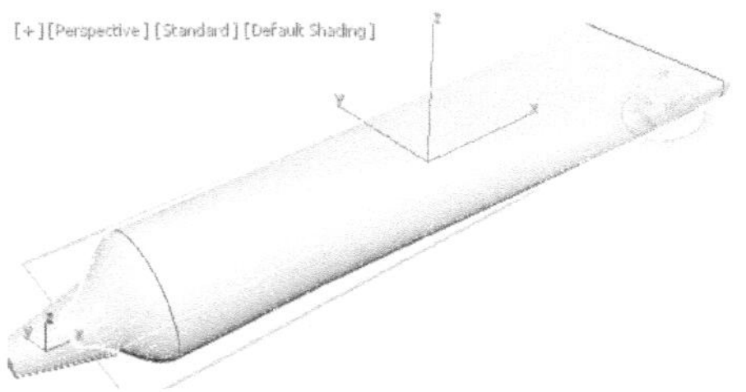

Figure 10-31 *The map on the toothpaste tube*

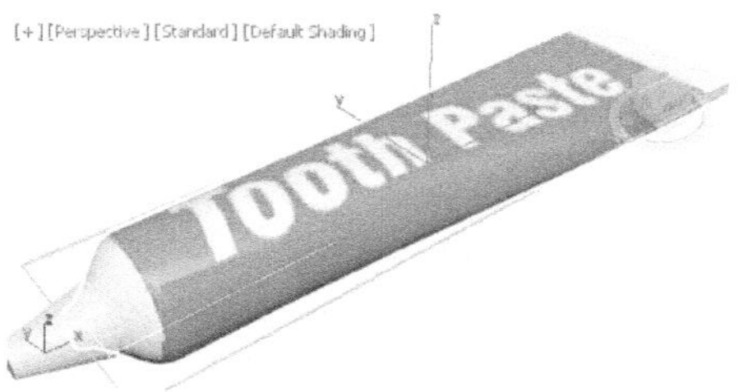

Figure 10-32 *The map on the toothpaste tube after applying the **UVW Map** modifier*

10. Select *cap01* in the viewport and select another sample slot in the **Material Editor** dialog box.

11. By default, the **Standard** material is selected in the **Material Editor** dialog box and you will use the same for *cap01*. Make sure that the **Blinn** shader is selected in the **Shader Basic Parameters** rollout. In the **Blinn Basic Parameters** rollout, choose the **Diffuse** color swatch; the **Color Selector: Diffuse Color** dialog box is displayed. Select the white color and choose the **OK** button to close the dialog box.

12. In the **Specular Highlights** area, set the following parameters:

    Specular Level: **44** Glossiness: **16**

13. Make sure that *cap01* is selected in the viewport. Next, choose the **Assign Material to Selection** tool in the **Material Editor** dialog box. Also, choose the **Show Shaded Material in Viewport** tool to display the map in the viewport. The toothpaste tube is displayed, refer to Figure 10-33. Close the **Material Editor** dialog box.

Figure 10-33 *The toothpaste tube after the map assigned*

Next, you need to apply the twist deformation.

14. Select *toothpaste tube* in the viewport and choose the **Modify** tab in the **Command Panel**.

15. In the modifier stack, select **Loft** and expand the **Deformations** rollout in the Modify panel. Now, in the **Deformations** rollout, choose the **Twist** button; the **Twist Deformation** dialog box is displayed, as shown in Figure 10-34.

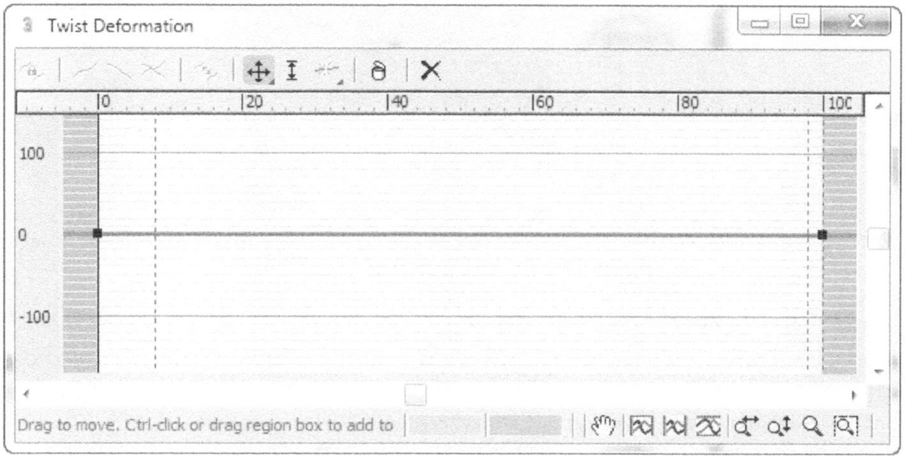

*Figure 10-34 The **Twist Deformation** dialog box*

16. In the **Twist Deformation** dialog box, choose the **Move Control Point** tool. Next, move the first vertex point (placed on the left side) vertically downward to **-65.0**. Alternatively, you can set the exact value in the edit box given at the bottom of this dialog box, refer to Figure 10-35; the toothpaste tube geometry gets twisted, as shown in Figure 10-36. Close the **Twist Deformation** dialog box.

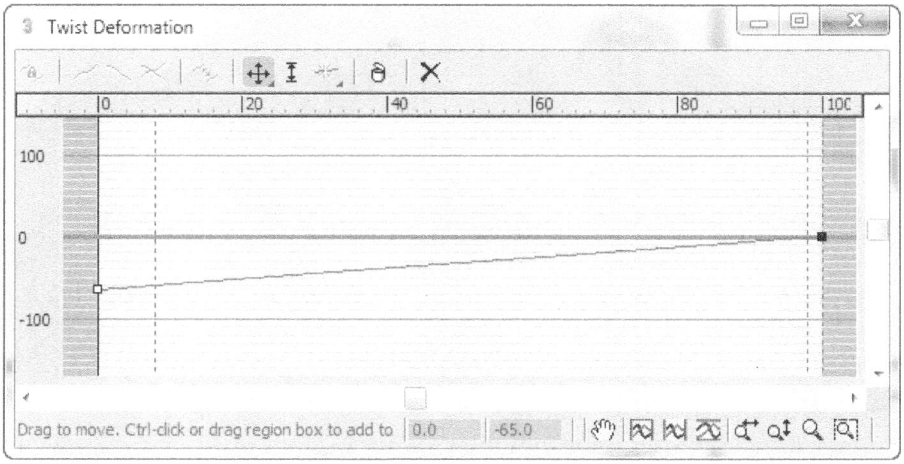

*Figure 10-35 The first vertex point moved in the **Twist Deformation** dialog box*

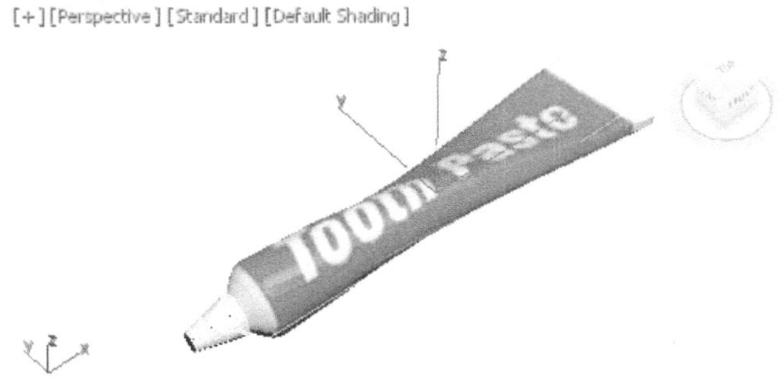

Figure 10-36 *Twisted toothpaste tube*

Saving and Rendering the Scene

In this section, you will save the scene and then render it. You can also view the final rendered image of this model by downloading the file *c10_3dsmax_2017_rndr.zip* from *www.www.cadsofttech.com*.The path of the file is as follows: *Textbooks > Animation and Visual Effects > 3ds Max > Autodesk 3ds Max 2017 for Novices*.

1. Choose **Save** from the **Application** menu. Activate the Perspective viewport. Render the scene using the **Render Production** tool in the **Main Toolbar**; the final output is displayed, refer to Figures 10-37 and 10-38.

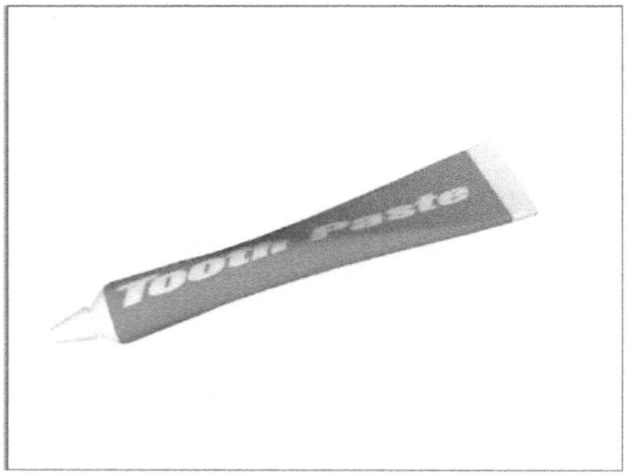

Figure 10-37 *The final output after rendering (view 1)*

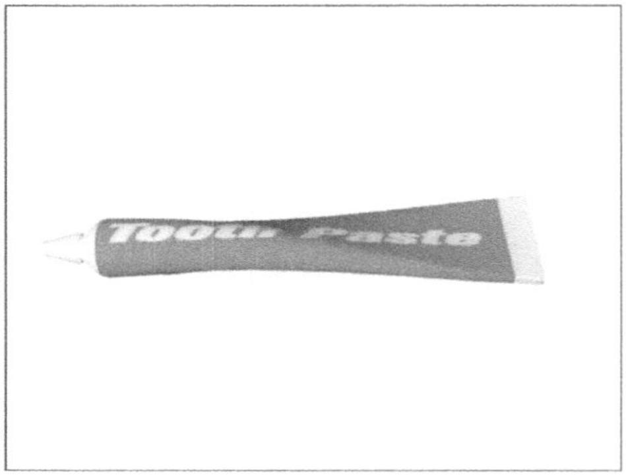

Figure 10-38 *The final output after rendering (view 2)*

Saving and Rendering the Scene

In this section, you will save the scene and then render it. You can also view the final rendered image of this model by downloading the file *c10_3dsmax_2017_rndr.zip* from.*www.cadsofttech.com*. The path of the file is as follows: *Textbooks > Animation and Visual Effects > 3ds Max > Autodesk 3ds Max 2017 for Novices*.

1. Activate the Perspective viewport. Next, choose **Save** from the **Application** menu.

2. Choose the **Render Production** tool in the **Main Toolbar**; the **Rendered Frame** window is displayed with the final output of the scene at one frame. Next, close this window.

EXERCISES

The rendered output of the models used in the following exercises can be accessed by downloading the *c10_3dsmax_2017_exr.zip* file from *www.cadsofttech.com*.The path of the file is as follows: *Textbooks > Animation and Visual Effects > 3ds Max > Autodesk 3ds Max 2017 for Novices.*

Exercise 1

Create the model shown in Figure 10-39 using the **ShapeMerge** tool in the **Compound Objects**. **(Expected time: 15 min)**

Exercise 2

Create the model shown in Figure 10-40 using the **Loft** tool in the **Compound Objects**. **(Expected time: 20 min)**

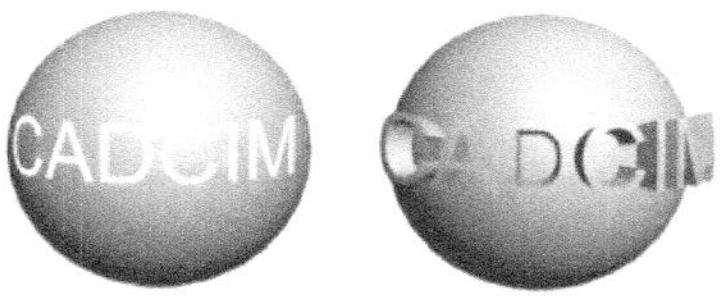

Figure 10-39 *Model to be created in Exercise 1*

Figure 10-40 *Model to be created in Exercise 2*

This page is intentionally left blank

Chapter 11

Modifiers

Learning Objectives

After completing this chapter, you will be able to:

- *Use modifiers*
- *Understand the types of modifiers*
- *Create complex objects using modifiers*

INTRODUCTION

In this chapter, you will learn about modifiers. As the name indicates, modifiers are used to modify an object. They enable you to create models that would otherwise be quite difficult to create.

MODIFIERS AND MODIFIER STACK

In 3ds Max, the modifiers are used to modify an object. You can apply more than one modifier to an object. The modifiers can also be used to modify the object at sub-object levels.

To apply a modifier to an object, select the object in the viewport and then choose the **Modify** tab in the **Command Panel**; the **Modifier List** drop-down list will be displayed, as shown in Figure 11-1. To view the options in the drop-down list, click on the arrow on the right of **Modifier List**, as shown in Figure 11-2. There are different types of modifiers in this list. Select the modifier that you want to apply to the object; the selected modifier will be displayed in the modifier stack along with various rollouts in the **Modify** panel to modify the object, refer to Figure 11-3.

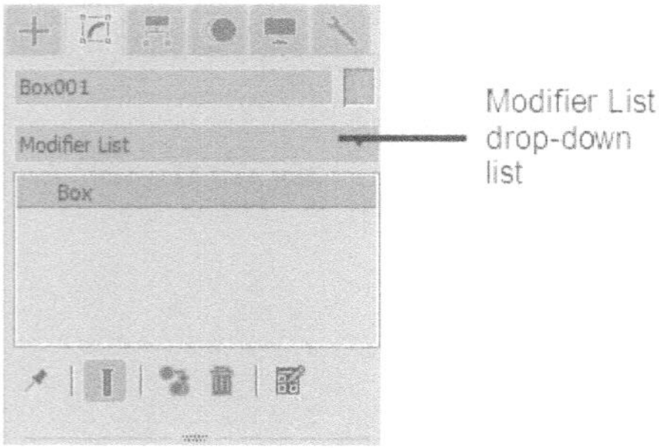

Figure 11-1 The ***Modifier List*** *drop-down list*

Figure 11-2 Partial view of the ***Modifier List*** *drop-down list*

Note

The modifiers are displayed in the ***Modifier List*** *drop-down list based on the selection of the object in the viewport.*

In the modifier stack, there is an eye icon, refer to Figure 11-3. It is used to show or hide the effects of the modifier in the viewport. By default, this icon is active. Click on this icon to make it inactive; the button will become dark and the effect of the modifier will not be displayed in the viewport. The buttons at the bottom of the modifier stack are used to manage the modifier stack.

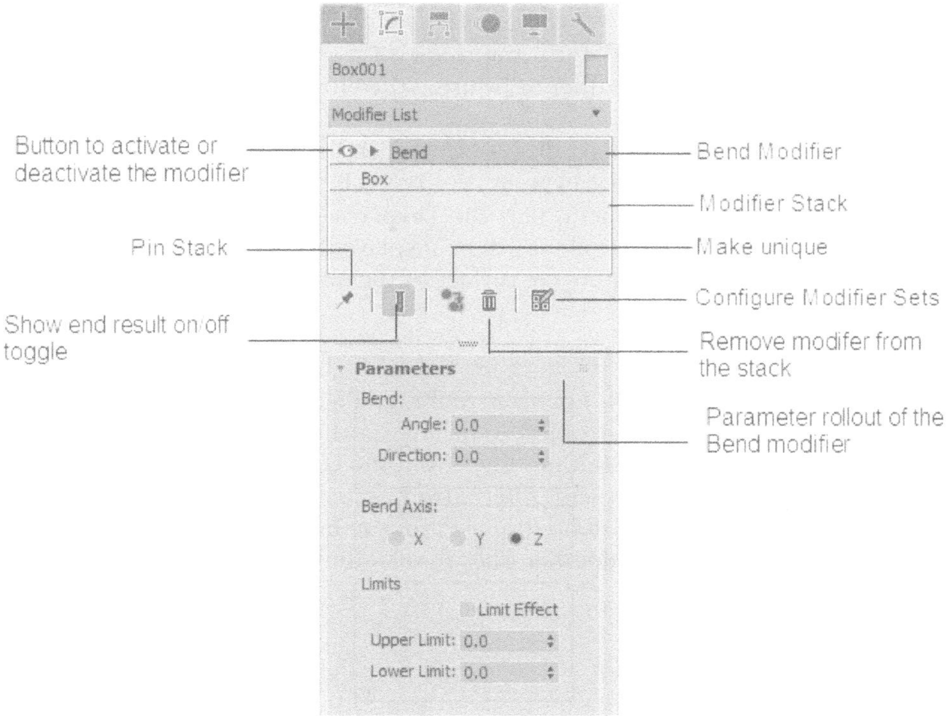

Figure 11-3 The modifier stack

TYPES OF MODIFIERS

There are three categories of modifiers in the **Modifier List** drop-down list: **Selection Modifiers**, **WORLD-SPACE MODIFIERS**, and **OBJECT-SPACE MODIFIERS**.

Extrude Modifier

The **Extrude** modifier is used to convert a 2D spline into a 3D object. It provides thickness and depth to a spline. To apply the **Extrude** modifier, select the spline in the viewport and choose the **Modify** tab in the **Command Panel**. Now, select the **Extrude** modifier from the **OBJECT-SPACE MODIFIERS** category in the **Modifier List** drop-down list. Alternatively, choose **Modifiers > Mesh Editing > Extrude** from the menu bar; the modifier will be displayed in the modifier stack and the **Parameters** rollout will be displayed below the modifier stack in the **Modify** panel, as shown in Figure 11-4.

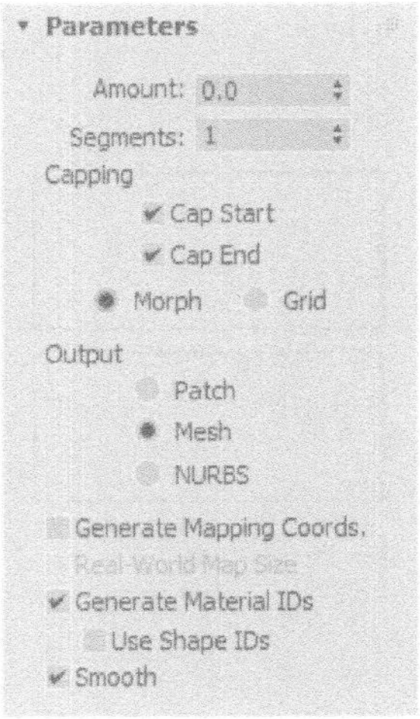

***Figure 11-4** The **Parameters** rollout*

Displace Modifier

The **Displace** modifier is used to deform the shape of an object. You can distort the object either by using the gizmo of the modifier or by using the bitmap image. To apply the **Displace** modifier, select the object in the viewport and choose the **Modify** tab in the **Command Panel**. Now, click on the **Modifier List** drop-down list and select the **Displace** modifier from the **OBJECT-SPACE MODIFIERS** category. Alternatively, choose **Modifiers > Parametric Deformers > Displace** from the menu bar; the **Displace** modifier will be displayed in the modifier stack and the **Parameters** rollout will be displayed in the Modify panel. You need to use the options in this rollout to distort the surface of the object.

Melt Modifier

The **Melt** modifier is used to apply a realistic melting effect to all types of objects. To apply the **Melt** modifier, select the editable object in the viewport and choose the **Modify** tab in the **Command Panel**. Now, select the **Melt** modifier from the **OBJECT-SPACE MODIFIERS** category in the **Modifier List** drop-down list. Alternatively, choose **Modifiers > Animation > Melt** from the menu bar; the modifier will be displayed in the modifier stack and the **Parameters** rollout will be displayed below the modifier stack in the Modify panel, as shown in Figure 11-5.

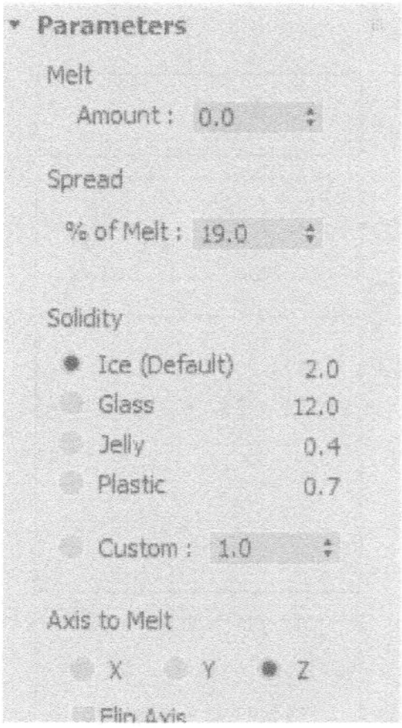

Figure 11-5 *The* ***Parameters*** *rollout*

Lattice Modifier

The **Lattice** modifier is used to convert the segments or edges of an object or shape into wireframes or cylindrical struts. To apply the **Lattice** modifier, select the object or shape in the viewport and choose the **Modify** tab in the **Command Panel**. Now, select the **Lattice** modifier from the **OBJECT-SPACE MODIFIERS** category in the **Modifier List** drop-down list. Alternatively, choose **Modifiers > Parametric Deformers > Lattice** from the menu bar; the modifier will be displayed in the modifier stack and its effect will be displayed on the object, as shown in Figure 11-6. The **Parameters** rollout will also be displayed below the modifier stack, as shown in Figure 11-7. Now, you need to use the options in the **Parameters** rollout to modify the effects of the **Lattice** modifier.

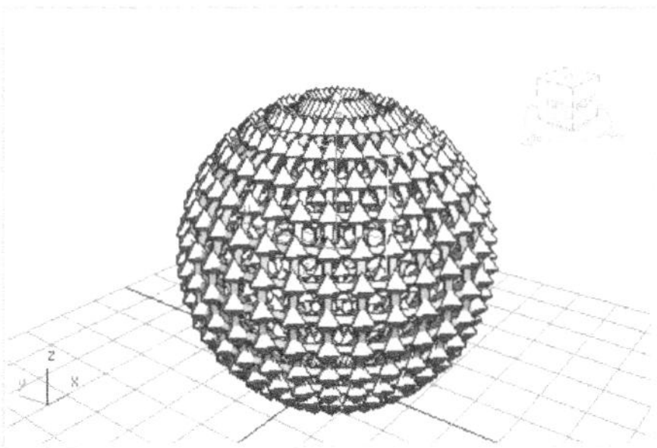

***Figure 11-6** The sphere after applying the **Lattice** modifier*

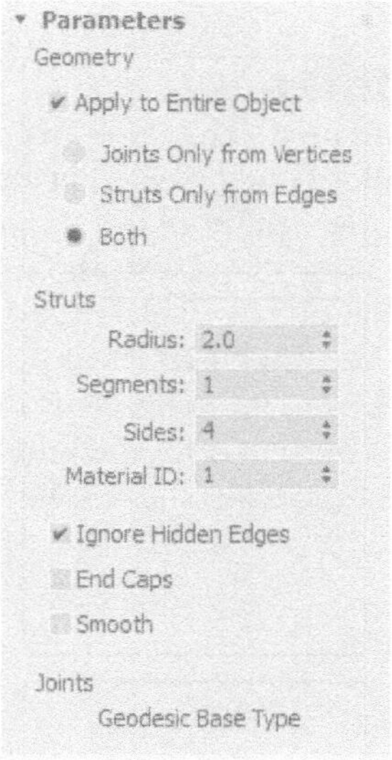

***Figure 11-7** Partial view of the **Parameters** rollout*

Noise Modifier

The **Noise** modifier is used to generate disturbances on the surface of an object to create irregular surfaces. It can also be used to animate water or wavy surfaces.

To apply the **Noise** modifier, select the object in the viewport and choose the **Modify** tab in the **Command Panel**. Now, select the **Noise** modifier from the **OBJECT-SPACE MODIFIERS** category in the **Modifier List** drop-down list. You can also choose **Modifiers > Parametric Deformers > Noise** from the menu bar; the **Noise** modifier will be displayed in the modifier stack and the **Parameters** rollout will also be displayed in the Modify panel. Next, you need to use the options in the **Parameters** rollout, refer to Figure 11-8, to apply the effects of the **Noise** modifier on the selected object.

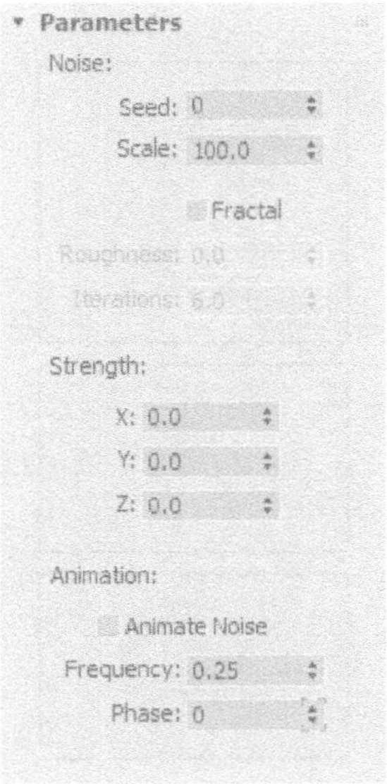

***Figure 11-8** The **Parameters** rollout*

Twist Modifier

The **Twist** modifier is used to produce the swirling effect on an object. To apply the **Twist** modifier, select an object in the viewport. Select the **Twist** modifier from the **OBJECT-SPACE MODIFIERS** category in the **Modifier List** drop-down list. Alternatively, choose the **Modifiers > Parametric Deformers > Twist** from the menu bar; the **Twist** modifier will be displayed in the modifier stack and the **Parameters** rollout will also be displayed in the Modify panel. Next, you need to use the options in the **Parameters** rollout, refer to Figure 11-9, to apply

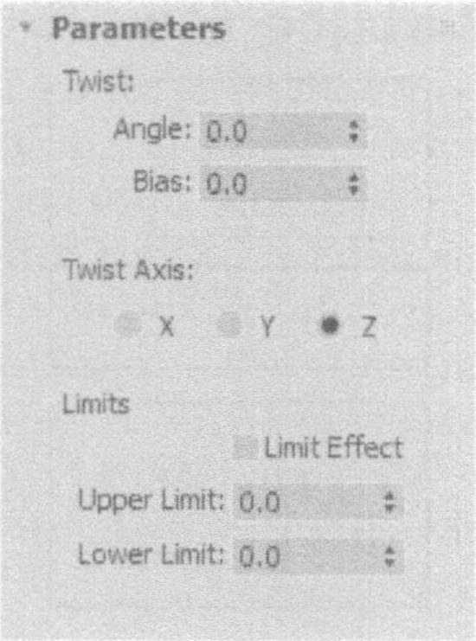

*Figure 11-9 The **Parameters** rollout*

Lathe Modifier

The **Lathe** modifier is a shape modifier that is used to create a 3D object by rotating a shape about an axis. The direction of the axis of revolution depends on the pivot point of the shape. To apply the **Lathe** modifier on an object, create a shape with the spline in the Front viewport and choose the **Modify** tab in the **Command Panel**. Now, select the **Lathe** modifier from the **OBJECT-SPACE MODIFIERS** category in the **Modifier List** drop-down list. Alternatively, choose **Modifiers > Patch/Spline Editing > Lathe** from the menu bar; the 2D shape will be converted into a 3D object, as shown in Figures 11-10 and 11-11. And, the modifier will be displayed in the modifier stack. The **Parameters** rollout will also be displayed below the modifier stack, refer to Figure 11-12.

*Figure 11-10 The **Line** spline*

*Figure 11-11 The **Line** spline after applying the **Lathe** modifier*

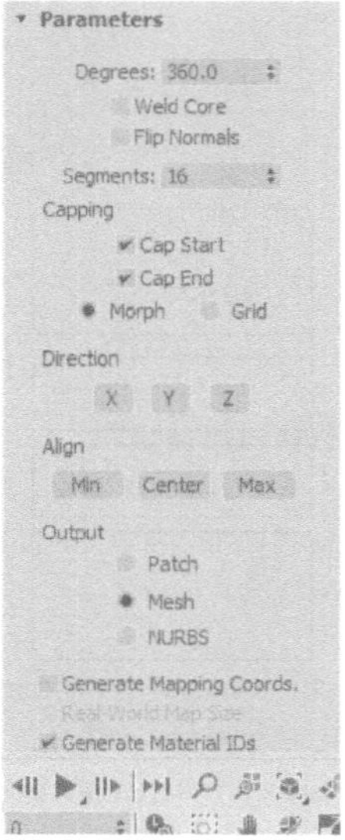

*Figure 11-12 Partial view of the **Parameters** rollout*

UVW Map Modifier

The **UVW Map** modifier is used to adjust the mapping coordinates on an object. Sometimes when you apply maps to objects, they are not displayed properly on its surface. To apply the map properly, you need to adjust the mapping coordinates on the object. The UVW coordinate system is the same as the XYZ coordinate system. The U, V, and W axes of a map correspond to the X, Y, and Z axes, respectively.

To apply the **UVW Map** modifier, select the object in the viewport and then select the **UVW Map** modifier from the **OBJECT-SPACE MODIFIERS** category in the **Modifier List** drop-down list. Alternatively, choose **Modifiers > UV Coordinates > UVW Map** from the menu bar; the **UVW Mapping** will be displayed in the modifier stack and the **Parameters** rollout will also be displayed below the modifier stack.

TUTORIAL

Tutorial 1

In this tutorial, you will create the 3D model of a tennis racket, as shown in Figure 11-13, using the **Lattice** modifier and shape splines. **(Expected time: 30 min)**

Figure 11-13 *A tennis racket model*

The following steps are required to complete this tutorial:

a. Create the project folder.
b. Create the head of the racket.
c. Create the shaft of the racket.
d. Create the handle of the racket.
e. Save and render the scene.

Creating the Project Folder

Create a new project folder with the name *c11_tut1* at *\Documents\3dsmax2017* and then save the file with the name *c11tut1*, as discussed in Tutorial 1 of Chapter 2.

Creating the Head of the Racket

In this section, you will create the head of the racket by using the **Ellipse** tool.

1. Choose **Create > Shapes** in the **Command Panel**; the **Splines** option is displayed in the drop-down list. Choose the **Ellipse** tool from the **Object Type** rollout.

2. Activate the Top viewport and create an ellipse. Choose the **Modify** tab in the **Command Panel**. Expand the **Rendering** rollout and select the **Enable In Renderer** and **Enable In Viewport** check boxes.

   Select the **Rectangular** radio button and enter the values given next:

   Length: **1.0** Width: **3.0**

   Make sure that the **Auto Smooth** check box is selected.
3. In the **Parameters** rollout, set the parameters as follows:

   Length: **145.894** Width: **93.72**

4. Modify the name of the ellipse to *head* and change its color to black.

   Next, you need to create the stringed area inside *head*.

5. Choose **Create > Geometry** in the **Command Panel**. Next, choose the **Plane** tool from **Standard Primitives** and create a plane in the Top viewport. In the **Parameters** rollout, set the following parameters:

   Length: **148.314** Width: **131.401**
   Length Segs: **12** Width Segs: **12**

6. Modify the name of the plane to *stringed area* and change its color to black. Now, align *stringed area* at the center of *head*, refer to Figure 11-14.

   Next, you need to modify *stringed area*.

7. Choose the **Maximize Viewport Toggle** tool to maximize the Top viewport.

8. Select *stringed area* in the viewport and choose the **Modify** tab in the **Command Panel**. Next, select the **Edit Mesh** modifier from the **OBJECT-SPACE MODIFIERS** category in the **Modifier List** drop-down list; the **Edit Mesh** modifier is displayed in the modifier stack and different rollouts are also displayed below the modifier stack.

Note

*By applying the **Edit Mesh** modifier, the name of the **Plane** tool will be displayed in the modifier stack. To modify creation parameters of the plane, you can select it from the modifier stack and set new values in the rollouts displayed in the Modify panel.*

9. In the **Selection** rollout, choose the **Vertex** button; all the vertices of *stringed area* are displayed in the viewport, as shown in Figure 11-15.

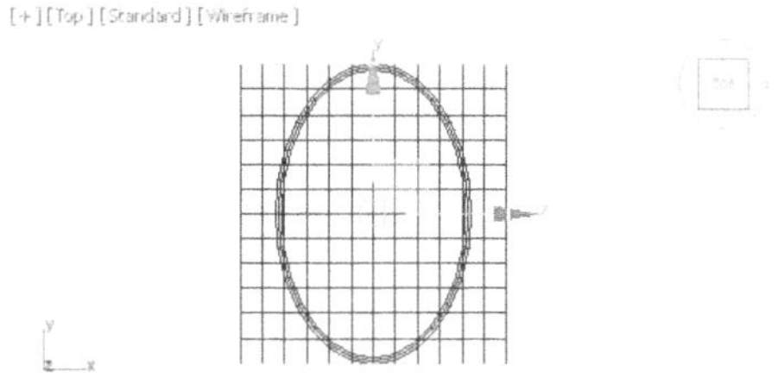

Figure 11-14 *The plane created in the Top viewport*

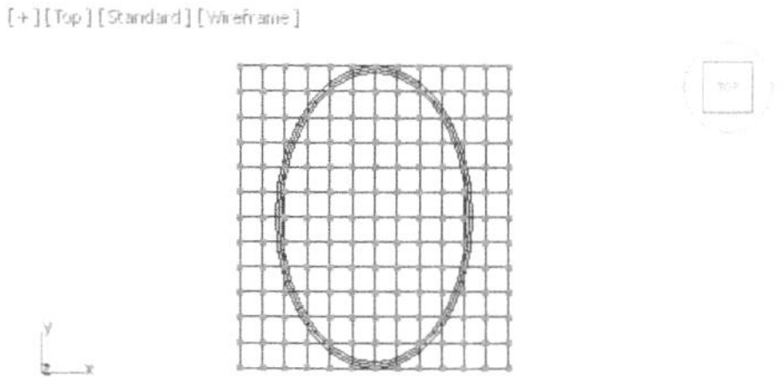

Figure 11-15 *The vertices of the stringed area displayed in the Top viewport*

Next, you need to select and move the vertices of *stringed area* one by one and then arrange them to form the shape of the head.

10. Move the vertices to the exact location by choosing the **Zoom Region** tool and drag a selection box around the vertices of the upper left portion of *stringed area* to zoom in on them, refer to Figures 11-16 and 11-17.

11. Choose the **Select and Move** tool to select the vertices one by one and arrange them to form the shape of *head*, as shown in Figure 11-18.

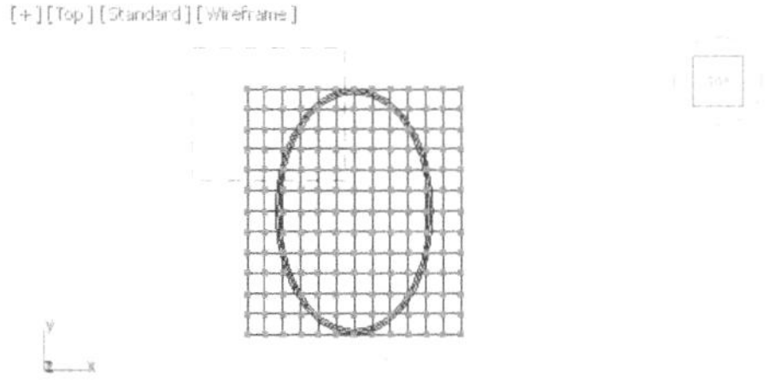

Figure 11-16 *A selection box dragged around the vertices using the **Zoom Region** tool*

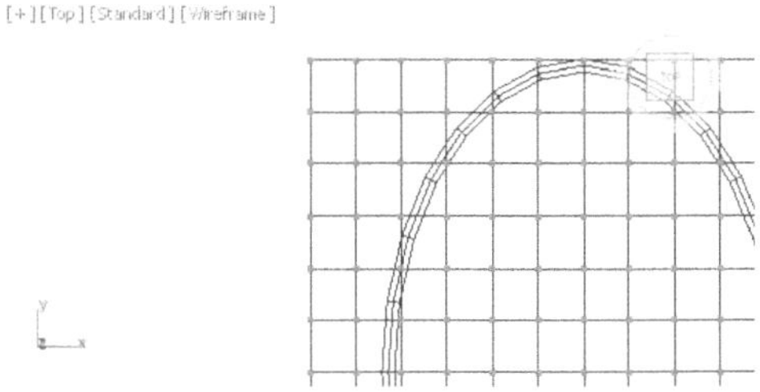

Figure 11-17 *The selected area zoomed in*

12. Choose the **Zoom Extents** tool; *head* and *stringed area* are displayed in the active viewport, refer to Figure 11-19.

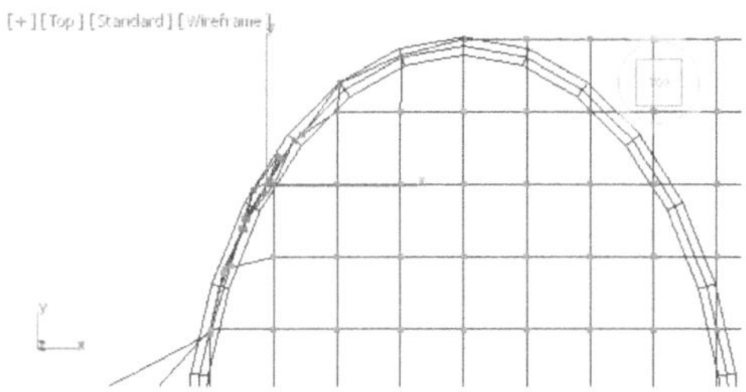

Figure 11-18 *The vertices arranged to form the shape of the head*

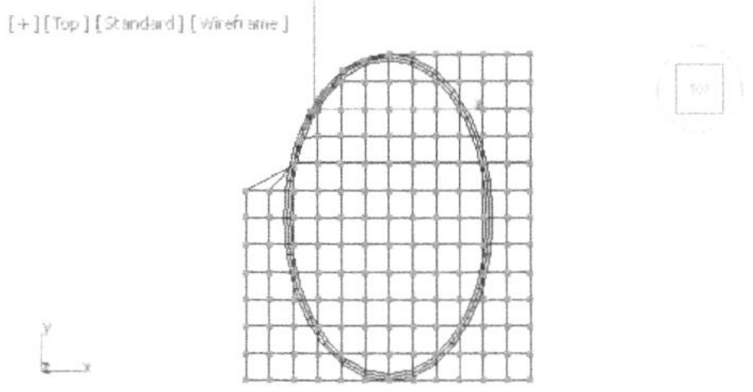

Figure 11-19 *The objects after invoking the **Zoom Extents** tool*

13. Follow the procedure discussed in steps 10 and 11 for the outer vertices and arrange them, as shown in Figure 11-20. Choose the **Vertex** button again in the **Selection** rollout to exit the sub-object level.

Next, you need to apply the **Lattice** modifier to *stringed area* of *head*.

14. Make sure that *stringed area* is selected in the viewport. Select the **Lattice** modifier from the **OBJECT-SPACE MODIFIERS** category in the **Modifier List** drop-down list; the **Lattice** modifier is displayed in the modifier stack and rollouts are also displayed below the modifier stack.

15. In the **Parameters** rollout, set the following parameters:

    Select the **Struts Only from Edges** radio button.

    Struts area
    Radius: **0.2**

    Stringed area is displayed, as shown in Figure 11-21.

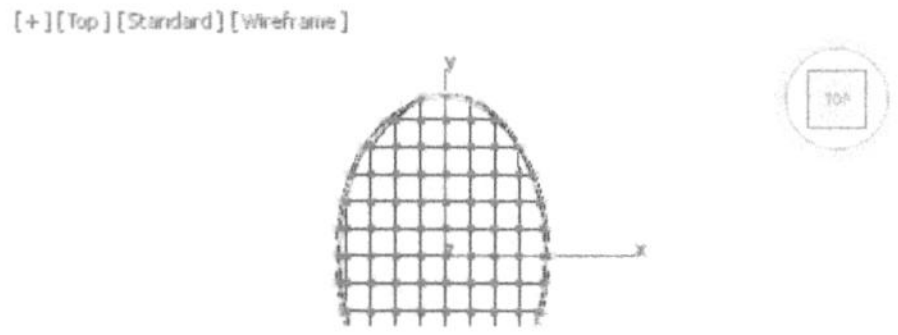

Figure 11-20 *All outer vertices arranged to form the shape of the head*

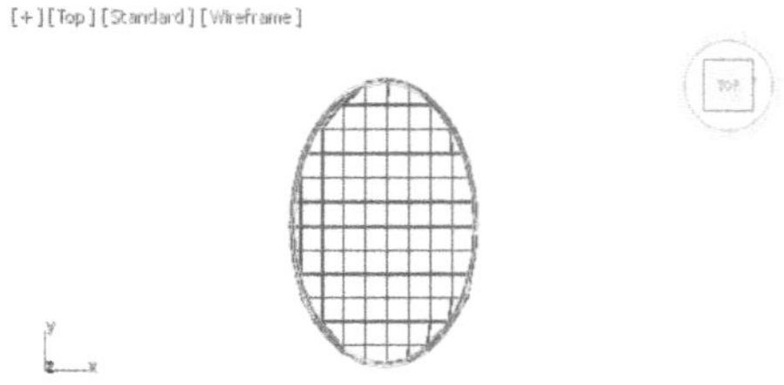

Figure 11-21 *The stringed area after applying the **Lattice** modifier*

16. Group *stringed area* and *head* as *racket head*.

Creating the Shaft of the Racket

In this section, you will create the shaft of the racket using the **Line** tool.

1. Choose the **Zoom** tool and zoom in the Top viewport to get sufficient room for creating the shaft of the racket, as shown in Figure 11-22.

2. Choose **Create > Shapes** in the **Command Panel**; the **Splines** option is displayed in the drop-down list. Choose the **Line** tool from the **Object Type** rollout.

3. In the **Creation Method** rollout, select the **Smooth** radio button in the **Initial Type** area and then select the **Smooth** radio button in the **Drag Type** area.

4. In the Top viewport, click and drag the cursor to create a line around *racket head*, as shown in Figure 11-23.

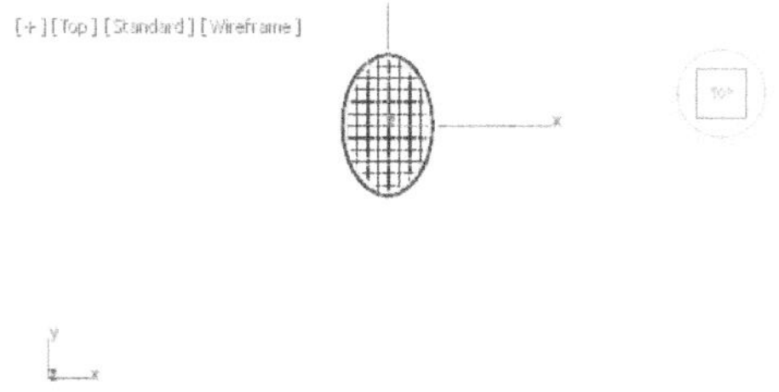

Figure 11-22 *The racket head in the Top viewport*

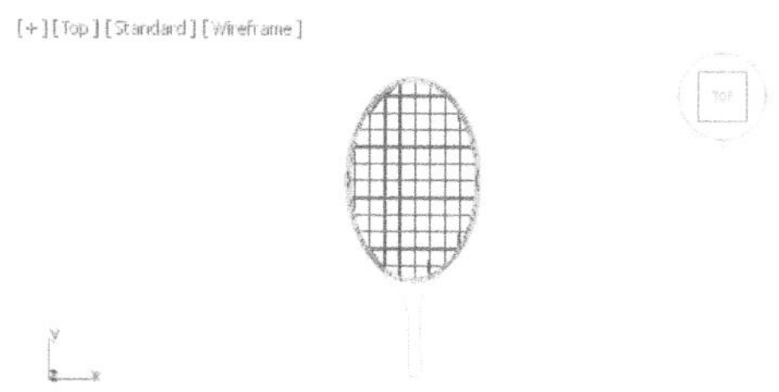

Figure 11-23 *A line created around the racket head*

5. In the **Rendering** rollout, make sure the **Enable In Renderer** and **Enable In Viewport** check boxes are selected. Also, select the **Radial** radio button and then enter the value **3** in the **Thickness** spinner.

6. Modify the name of the line to *shaft01* and then modify its color by setting the following parameters:

   Red: **8** Green: **8** Blue: **136**

7. Choose the **Maximize Viewport Toggle** tool and activate the Front viewport. Now, align *shaft01* at the bottom of *racket head*, refer to Figure 11-12.

8. In the Front viewport, select *shaft01* and create its copy in the vertical direction. It is automatically named as *shaft002*.

9. Align *shaft002* at the top of *racket head*, as shown in Figure 11-13.

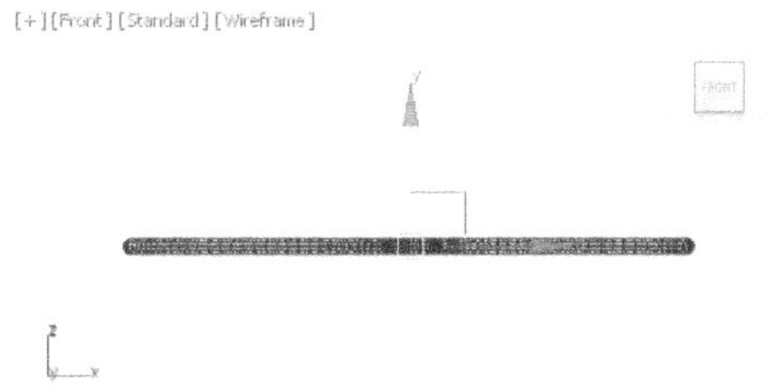

Figure 11-24 *Alignment of the shaft01 at the bottom of the racket head*

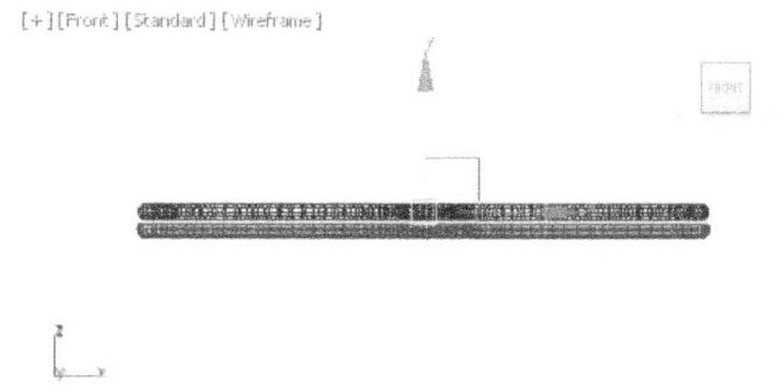

Figure 11-25 *Alignment of the shaft002 at the top of the racket head*

Creating the Handle of the Racket

In this section, you will create the handle of the racket by using the **ChamferCyl** tool from **Extended Primitives**.

1. Choose the **ChamferCyl** tool from **Extended Primitives** in the **Command Panel** and create a chamfer cylinder in the Front viewport.

2. In the **Parameters** rollout, set the following parameters:

Radius: **7.0** Height: **74.477** Fillet: **2.816**

3. Modify the name of the cylinder to *handle* and change its color to black.

4. Align *handle* with *shaft01* and *shaft002* in viewports, as shown in Figure 11-26.

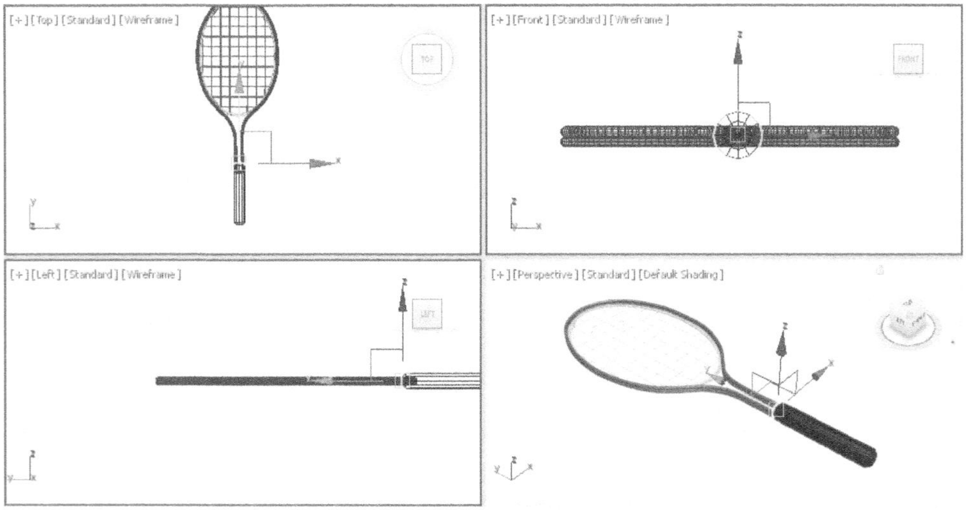

Figure 11-26 *Alignment of handle in viewports*

Saving and Rendering the Scene

In this section, you will save the scene and then render it. You can also view the final rendered image of this model by downloading the file *c11_3dsmax_2017_rndr.zip* from *www.cadsofttech.com*. The path of the file is as follows: *Textbooks > Animation and Visual Effects > 3ds Max > Autodesk 3ds Max 2017 for Novices*.

1. Change the background color of the scene to white as described in the previous chapters.

2. Choose **Save** from the **Application** menu. Next, activate the Perspective viewport.

3. Choose the **Render Production** tool from the **Main Toolbar**; the **Rendered Frame** window is displayed with the final output of the tennis racket, as shown in Figure 11-27. Next, close this window.

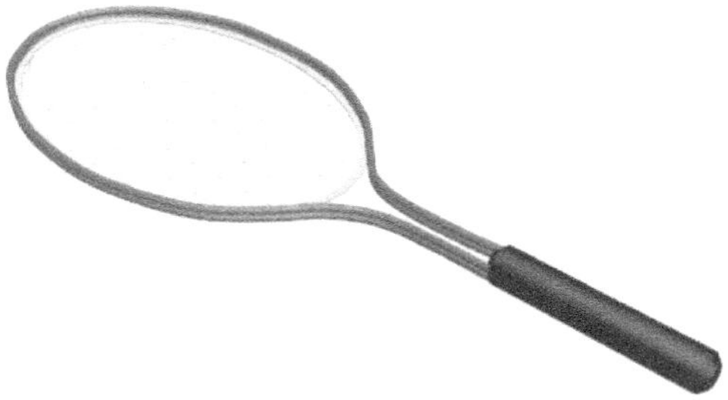

Figure 11-27 *The final output*

EXERCISE

The rendered output of the models used in the following exercises can be accessed by downloading the *c11_3dsmax_2017_exr.zip* file from *www.www.www.cadsofttech.com*.The path of the file is as follows: *Textbooks > Animation and Visual Effects > 3ds Max > Autodesk 3ds Max 2017 for Novices*.

Exercise 1

Create the model of a basket, as shown in Figure 11-28, using the **Lattice** modifier.

(Expected time: 15 min)

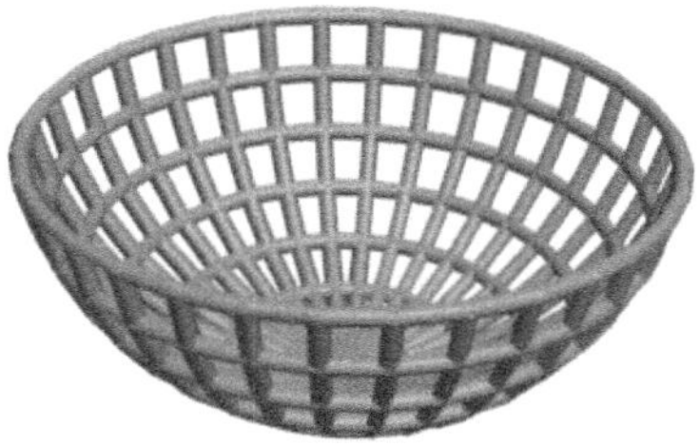

Figure 11-28 *The model of a basket*

This page is intentionally left blank

Chapter 12

Lights and Cameras

Learning Objectives

After completing this chapter, you will be able to:
- *Work with standard lights*
- *Work with photometric lights*
- *Work with Camera tools*

INTRODUCTION

In 3ds Max, lights are used to illuminate a scene to make it realistic. They are also used to create the sources of illumination such as street lights, flashlights, and so on. The two main categories of lights available in 3ds max are: **Standard** and **Photometric**. In these categories, different types of lights are available. You need to select one or more lights based on the requirement of a scene. The Photometric lights provide more realistic lighting effects but they are more complex than the Standard lights.

***Figure 12-1** The tools to create standard lights in the **Object Type** rollout*

STANDARD LIGHTS

In 3ds Max, the Standard lights are objects that simulate lights. You can create eight types of Standard lights in 3ds Max. To create a light in a scene, choose **Create > Lights** in the **Command Panel**. By default, the **Photometric** option will be displayed in the drop-down list below the **Lights** button. Select the **Standard** option from the drop-down list. You can choose one of the tools available in the **Object Type** rollout to create different types of lights in the viewport, refer to Figure 12-1. When you choose a tool from the **Object Type** rollout, various rollouts will be displayed. These rollouts are used to modify the parameters of lights.

PHOTOMETRIC LIGHTS

The photometric lights are used to add realistic light effects to the scenes. The photometric lights use the light energy values to calculate the lights accurately as they are in the real world. To create a photometric light in a scene, choose **Create > Lights** in the **Command Panel**; the **Photometric** light will be displayed by default in the drop-down list below the **Lights** button. Also, various tools to create photometric lights will be displayed in the **Object Type** rollout. These tools are **Target Light**, **Free Light**, **mr Sky Portal**, and **Sun Positioner**, refer to Figure 12-2. Choose one of these tools and create the light in the viewport. If you choose the **Target Light** or **Free Light** tool and if you are using the **Default Scanline Renderer**, the **Photometric Light Creation** message box will be displayed, asking if you want to use the Logarithmic Exposure Control, refer to Figure 12-3. Choose the **Yes** button in this message box. When you choose a tool from the **Object Type** rollout, various rollouts will be displayed to modify the parameters of the corresponding light. Note that if you choose the **Target Light** tool or **Free Light** tool and are using the **NVIDIA mental ray Renderer**, the **Photometric Light Creation** message box will be displayed, asking if you want to use the mr Photographic Exposure Control, refer to Figure 12-4.

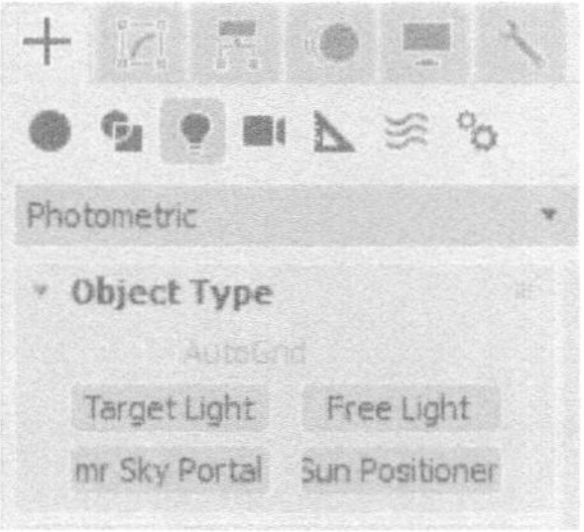

***Figure 12-2** Various tools to create **Photometric** lights*

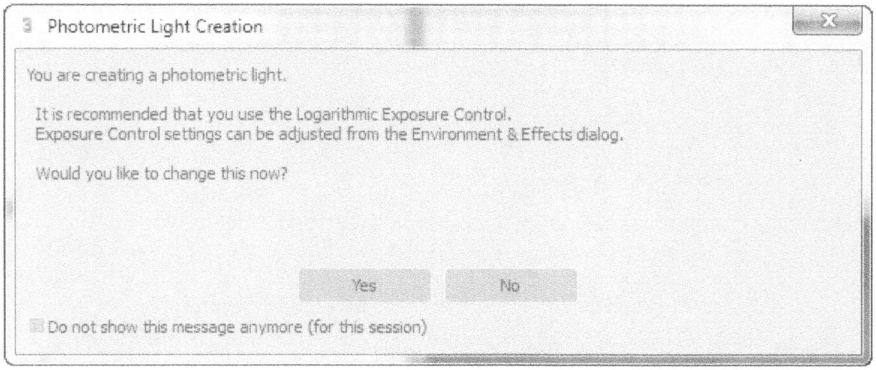

Figure 12-3 *The **Photometric Light Creation** message box*

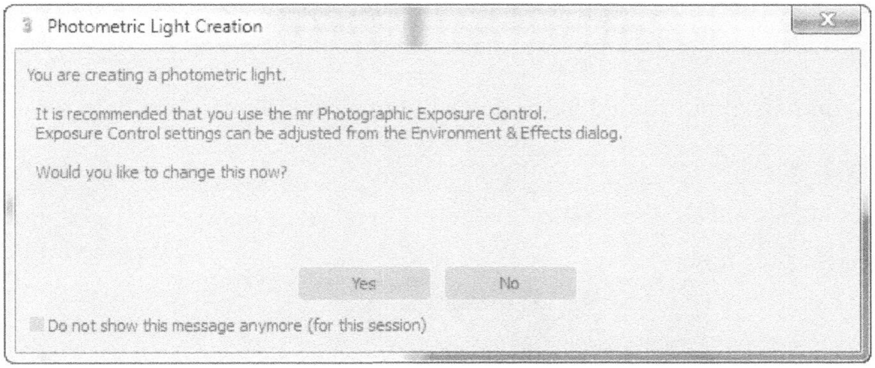

Figure 12-4 *The **Photometric Light Creation** message box*

The **Name and Color**, **General Parameters**, **Shadow Parameters**, **Advanced Effects**, and **Atmospheres & Effects** rollouts for photometric lights are same as discussed in the standard lights.

CAMERAS

In 3ds Max, the cameras are used to adjust a particular view in a scene. These cameras act as the still image or video cameras in the real world. There are three types of cameras in 3ds Max 2017, namely Target, Free and Physical. The target camera has two parts, the target camera and a target object, as shown in Figure 12-5. It displays the view according to the position of the target object. You can move the target object and the target camera independently. The free camera does not have any target object, refer to Figure 12-5. It displays the view in the direction of the camera. You can use the free camera to perform the walkthrough animation. The physical camera uses exposure control and produces photorealistic renderings.

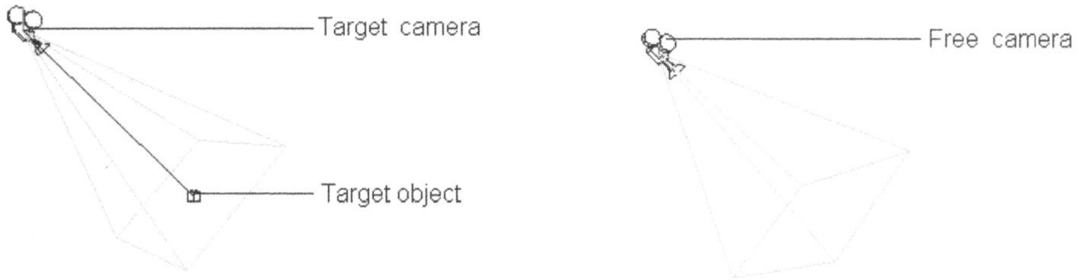

Figure 12-5 *The target and free cameras*

TUTORIALS

Before starting the tutorials, you need to download the *c12_3dsmax_2017_tut.zip* file from.*www.cadsofttech.com*.The path of the file is as follows: *Textbooks > Animation and Visual Effects > 3ds Max > Autodesk 3ds Max 2017 for Novices.*

Extract the contents of the zipped file and save them in the *Documents* folder.

Tutorial 1

In this tutorial, you will create a scene, as shown in Figure 12-6, by using the lights, materials and maps. **(Expected time: 30 min)**

Figure 12-6 *The scene to be created*

The following steps are required to complete this tutorial:

a. Create the project folder.
b. Open the file.
c. Create lights.
d. Save and render the scene.

Creating the Project Folder

1. Create a new project folder with the name *c12_tut1* at *\Documents\3dsmax2017* and then save the file with the name *c12tut1*, as discussed in Tutorial 1 of Chapter 2.

2. Open the Windows Explorer and then browse to the *c12_3dsmax_2017_tut* folder and copy the *sky_1.jpg* file from this folder to *\Documents\3dsmax2017\c12_tut1\sceneassets\images*.

Opening the File

1. Choose **Open** from the **Application** menu; the **Open File** dialog box is displayed. In this dialog box, browse to the location *\Documents\c12_3dsmax_2017_tut* and select the **c12_tut1_start.max** file from it. Choose the **Open** button to open the file, refer to Figure 12-7.

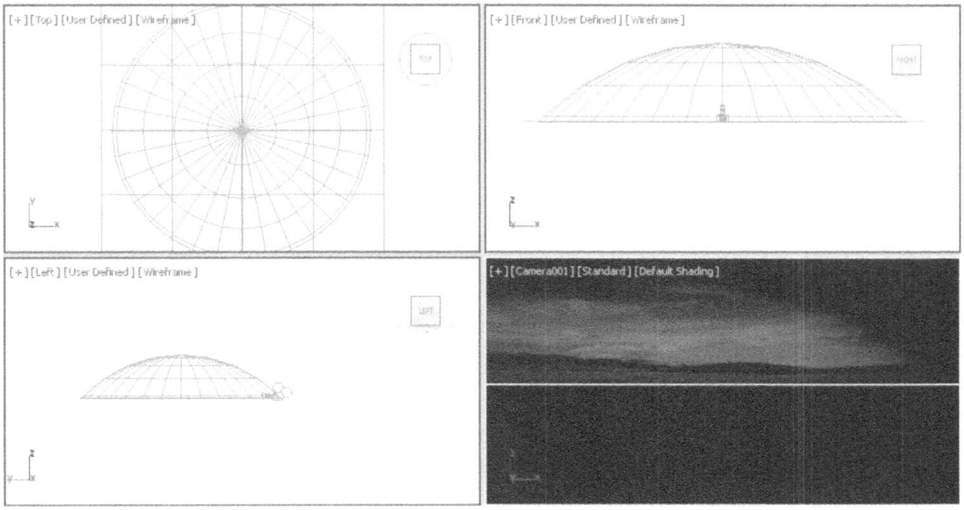

Figure 12-7 *The c12_tut1_start.max file*

2. Choose **Save As** from the **Application** menu; the **Save File As** dialog box is displayed. Browse to the location *\Documents\3dsmax2017\c12_tut1\scenes*. Save the file with the name *c12tut1.max* at this location.

Creating Lights

In this section, you will create lights by using the **Omni** tool.

1. Activate the Top viewport. Choose **Create > Lights** in the **Command Panel**; the **Photometric** option is displayed in the drop-down list below the **Lights** button. Choose the **Standard** option from the drop-down list.

2. Choose the **Omni** tool from the **Object Type** rollout in the **Command Panel** and click in the Top viewport; the omni light is displayed in the viewports. It is automatically named as *Omni001*.

3. Choose the **Select and Move** tool and align *Omni001* in the viewports, as shown in Figure 12-8.

4. Create one more omni light at the center of *background* in the Top viewport. It is automatically named as *Omni002*. Now, align *Omni002* in all viewports, as shown in Figure 12-9.

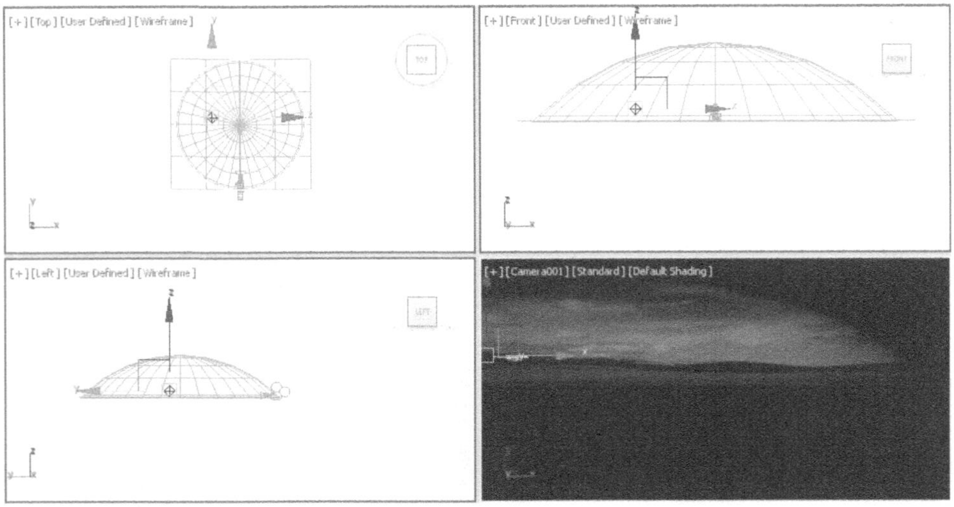

Figure 12-8 *Alignment of Omni001 in the viewports*

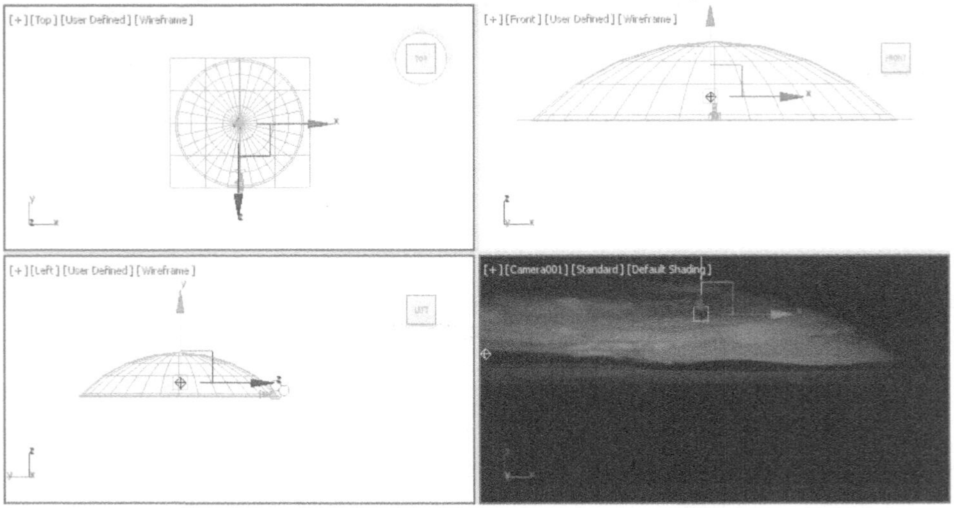

Figure 12-9 *Alignment of Omni002 in the viewports*

Next, you need to modify *Omni002* light to create the sun effect.

5. Make sure that *Omni002* light is selected and then choose the **Modify** tab in the **Command Panel**; the rollouts to modify the light are displayed in the modify panel.

6. Expand the **Atmospheres & Effects** rollout and choose the **Add** button; the **Add Atmosphere or Effect** dialog box is displayed. Select the **Lens Effects** option and then choose the **OK** button; the **Lens Effects** option is displayed in the text area of the **Atmospheres & Effects** rollout. Next, you need to modify the parameters of **Lens Effects**.

7. Select **Lens Effects** in the **Atmospheres & Effects** rollout and choose the **Setup** button; the **Environment and Effects** dialog box is displayed. By default, the **Effects** tab is chosen in the **Environment and Effects** dialog box.

8. In the **Effects** rollout, make sure the **Lens Effects** option is selected under the **Effects** area. Now, select the **Glow** effect and choose the right arrow button in the **Lens Effects Parameters** rollout; the **Glow** effect is displayed in the list available on the right side, as shown in Figure 12-10. Also, the **Glow Element** rollout is displayed at the bottom of the **Environment and Effects** dialog box to modify the parameters of the **Glow** effect.

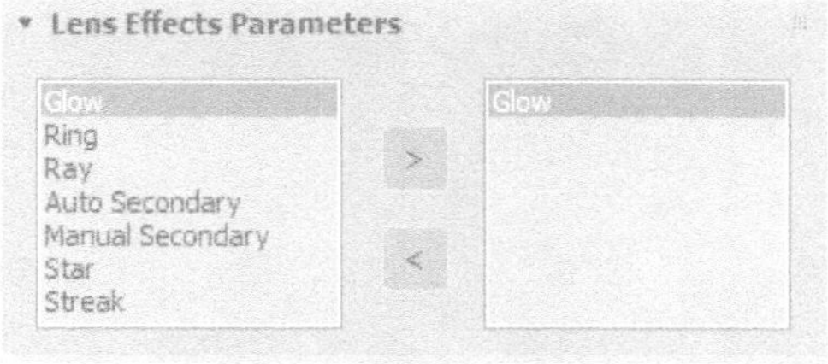

***Figure 12-10** The **Lens Effects Parameters** rollout*

Note

1. To view all rollouts, you need to scroll inside the dialog box.
*2. You can view the effects of the **Lens Effects** option only after rendering.*

9. In the **Glow Element** rollout, set the value **20.0** in the **Size** spinner. Use the default values for other options. In the **Lens Effects Parameters** rollout, select the **Ray** effect and choose the right arrow button; the **Ray** effect is displayed in the text area available on the right side. Also, the **Ray Element** rollout is displayed at the bottom of the **Environment and Effects** dialog box to modify the parameters of the **Ray** effect.

10. In the **Ray Element** rollout, set the values in the spinners as follows and close the **Environment and Effects** dialog box.

    Size: **50.0** Num: **20**

Saving and Rendering the Scene

In this section, you will save the scene and then render it. You can also view the final rendered image of this model by downloading the file *c12_3dsmax_2017_rndr.zip* from *www.cadsofttech.com*. The path of the file is as follows: *Textbooks > Animation and Visual Effects > 3ds Max > Autodesk 3ds Max 2017 for Novices*.

1. Choose **Save** from the **Application** menu. Next, activate the Camera001 viewport.

2. Choose the **Render Production** tool; the **Rendered Frame** window is displayed showing the final output of the scene at one frame, as shown in Figure 12-11.

Figure 12-11 *The final output of the scene*

3. Close the **Rendered Frame** window.

Tutorial 2

In this tutorial, you will illuminate a room by using the **Omni** and **Target Direct** tools, as shown in Figure 12-12. **(Expected time: 30 min)**

Figure 12-12 *The illuminated room*

The following steps are required to complete this tutorial:

a. Create the project folder.
b. Open the file.
c. Create lights.
d. Create the background.

e. Save and render the scene.

Creating the Project Folder

Create a new project folder with the name *c12_tut2* at *\Documents\3dsmax2017* and then save the file with the name *c12tut2*, as discussed in Tutorial 1 of Chapter 2. Now, copy *Finishes.Masonry Flooring.Marble.Beige-Grid.jpg* and *CLOUDS4.jpeg* files to *\Documents\3dsmax2017\c12_tut2\sceneassets\images*.

Opening the File

1. Choose **Open** from the **Application** menu; the **Open File** dialog box is displayed. In this dialog box, browse to the location *\Documents\c12_3dsmax_2017_tut* and select the **c12_tut2_start.max** file from it. Choose the **Open** button to open the file, refer to Figure 12-13.

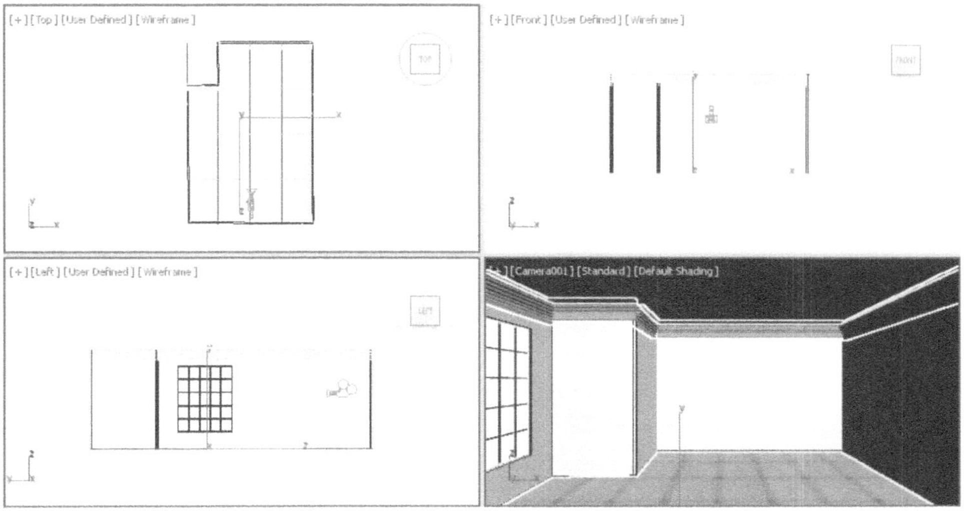

Figure 12-13 *The c12_tut2_start.max file*

2. Choose **Save As** from the **Application** menu; the **Save File As** dialog box is displayed. Browse to the location *\Documents\3dsmax2017\c12_tut2\scenes*. Save the file with the name *c12tut2.max* at this location.

Creating Lights

In this section, you will create lights by using the **Omni** and **Target Direct** tools.

1. Activate the Top viewport. Choose **Create > Lights** in the **Command Panel** and select the **Standard** option from the drop-down list. Next, choose the **Omni** tool from the **Object Type** rollout. Next, click in the Top viewport; the omni light is displayed in the viewport and it is automatically named as *Omni001*.

2. In the **Modify** tab, set the following parameters:

   General Parameters rollout

Select the **On** check box in the **Shadows** area
Intensity/Color/Attenuation rollout
Set the value **0.6** in the **Multiplier** spinner

3. Select the **Multiplier** color swatch and modify the colors by setting the following parameters:

   Red: **255** Green: **219** Red: **181**

4. Align *Omni001* in viewports, as shown in Figure 12-14.

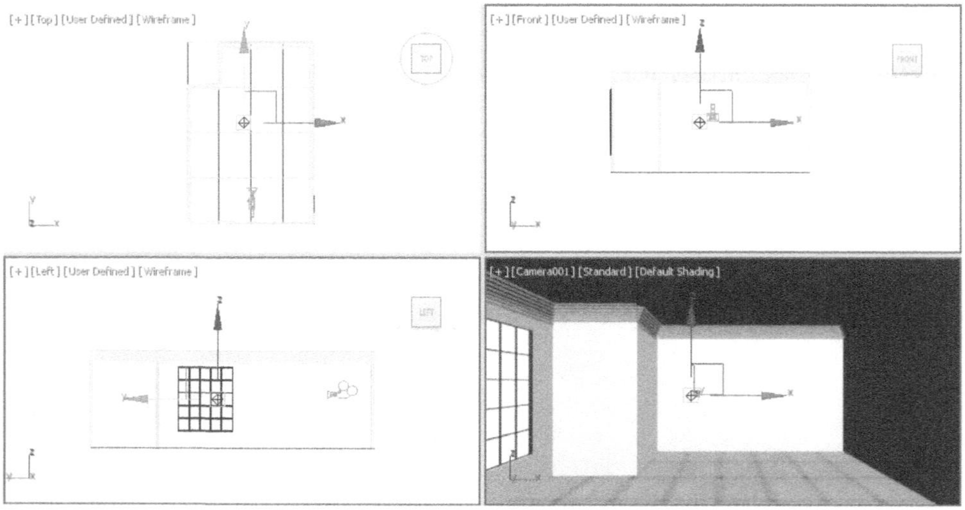

Figure 12-14 *The Omni001 light in viewports*

Next, you need to create the target direct light in the scene to give the shadow effect.

5. Activate the Front viewport, choose the **Zoom** tool, and zoom out until the objects are displayed about half of their original size.

6. Choose the **Target Direct** tool from **Create > Lights > Standard > Object Type** rollout in the **Command Panel**. Move the cursor on the left side of the Front viewport; the shape of the cursor changes. Next, press and hold the left mouse button to specify the location of the light and drag the cursor near the front portion of *wall001* to specify the target of the light. Release the left mouse button; the light is displayed in all viewports and it is automatically named as *Direct001*.

7. Choose the **Modify** tab in the **Command Panel**; various rollouts are displayed to modify the target direct light.

8. Expand the **General Parameters** rollout and select the **On** check box in the **Shadows** area.

9. In the **Directional Parameters** rollout, select the **Rectangle** radio button and set the values in the respective spinners as given below.

Hotspot/Beam: **84.0**   Falloff/Field: **86.0**

10. Expand the **Atmospheres & Effects** rollout and choose the **Add** button; the **Add Atmosphere or Effect** dialog box is displayed. Select the **Volume Light** option and choose the **OK** button; the **Volume Light** option is displayed in the text area of the **Atmospheres & Effects** rollout.

    Next, you need to modify the parameters of the **Volume Light**.

11. Select the **Volume Light** option in the **Atmospheres & Effects** rollout and choose the **Setup** button; the **Environment and Effects** dialog box is displayed. The **Environment** tab is chosen by default in the **Environment and Effects** dialog box.

12. In the **Atmosphere** rollout, make sure the **Volume Light** option is selected in the **Effects** area. Set the value **1.5** in the **Density** spinner of the **Volume** area in the **Volume Light Parameters** rollout. Use the default values for other options and close the **Environment and Effects** dialog box.

Note
You can also use different settings in the ***Volume Light Parameters*** *rollout to give different effects in the scene.*

13. Align *Direct001* light in the viewports to cover window, as shown in Figure 12-15.

Note
You can adjust the position of the Camera001 to view the scene with desired light effects.

Creating Background

In this section, you will create a background for the scene across the window.

1. Choose **Rendering > Environment** from the menu bar; the **Environment and Effects** dialog box is displayed.

2. The **Environment** tab is chosen by default in the **Environment and Effects** dialog box. In the **Background** area of the **Common Parameters** rollout, choose the **Environment Map** button that is currently labeled as **None**; the **Material/Map Browser** dialog box is displayed.

3. Select the **Bitmap** map and choose the **OK** button; the **Select Bitmap Image File** dialog box is displayed. As the project folder is already set, the *images* folder is displayed in the **Look in** drop-down list of this dialog box. Select the **CLOUDS4.jpg** file and choose the **Open** button; the selected image is displayed in the background after rendering. Close the **Environment and Effects** dialog box.

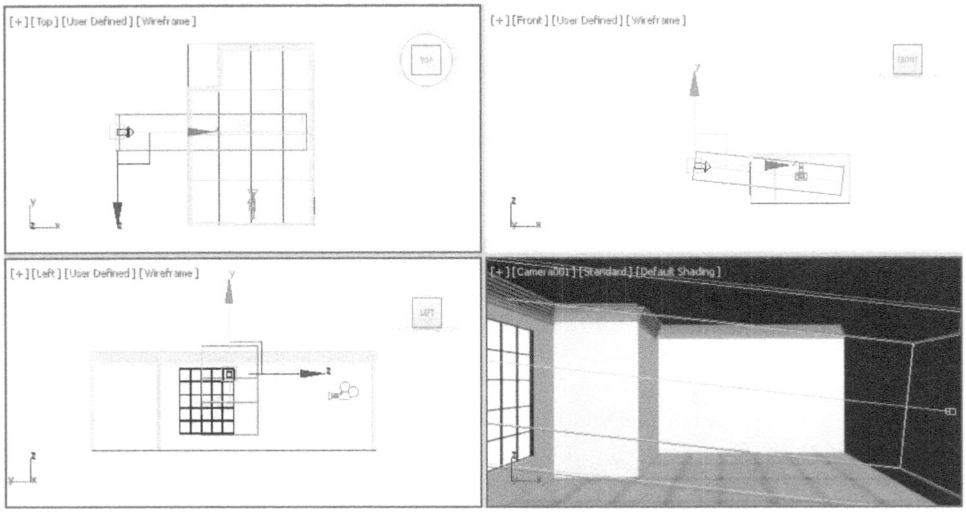

Figure 12-15 Alignment of the Direct001 light in viewports

Saving and Rendering the Scene

In this section, you will save the scene and then render it. You can also view the final rendered image of this model by downloading the file *c12_3dsmax_2017_rndr.zip* from *www.cadsofttech.com*. The path of the file is as follows: *Textbooks > Animation and Visual Effects > 3ds Max > Autodesk 3ds Max 2017 for Novices*.

1. Choose **Save** from the **Application** menu. Activate the Camera001 viewport.

2. Choose the **Render Production** tool in the **Main Toolbar**; the **Rendered Frame** window is displayed, showing the final output of the scene, as shown in Figure 12-16.

3. Close this window.

Figure 12-16 The final output after rendering

Tutorial 3

In this tutorial, you will illuminate a night scene of a room using photometric lights, refer to Figure 12-17. **(Expected time: 30 min)**

Figure 12-17 The illuminated night scene of a room

The following steps are required to complete this tutorial:

a. Create the project folder.
b. Open the file.
c. Create the roof light.
d. Create wall lights.
e. Save and render the scene.

Creating the Project Folder

1. Create a project folder with the name *c12_tut3* in the *3dsmax2017* folder, as discussed in Tutorial 1 of Chapter 2.

2. Open Windows Explorer and then browse to *\Documents\c12_3dsmax_2017_tut*. Next, copy all jpeg files to *\Documents\3dsmax2017\c12_tut3\sceneassets\images*. Also, copy *1.IES* file to *\Documents\3dsmax2017\c12_tut3\sceneassets\photometric*.

Opening the File

1. Choose **Open** from the **Application** menu; the **Open File** dialog box is displayed. In this dialog box, browse to the location *\Documents\c12_3dsmax_2017_tut* and select the **c12_tut3_start.max** file from it. Choose the **Open** button to open the file, refer to Figure 12-18.

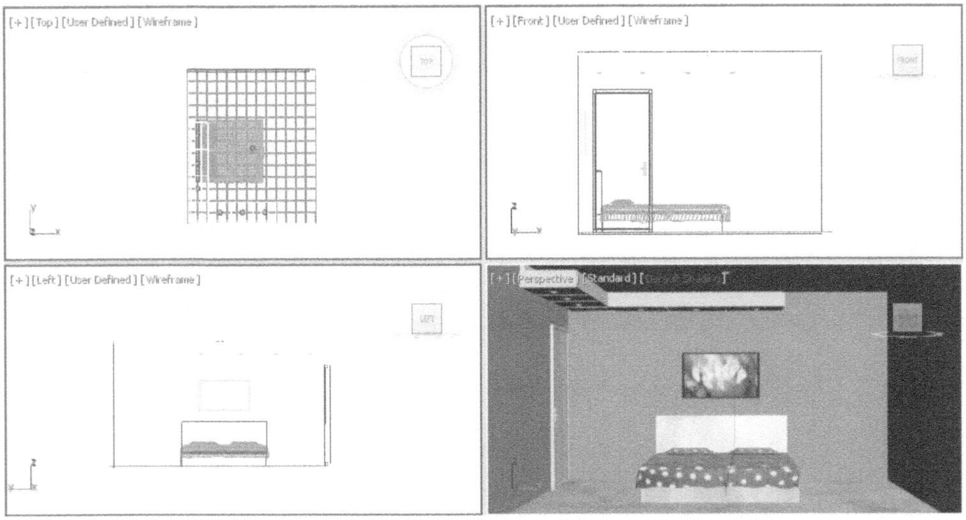

Figure 12-18 The c12_tut3_start.max file

2. Choose **Save As** from the **Application** menu; the **Save File As** dialog box is displayed. Browse to the location *\Documents\3dsmax2017\c12_tut3\scenes*. Save the file with the name *c12tut3.max* at this location.

Creating the Roof Light

In this section, you will create the roof light by using the **Free Light** tool. You will also change the renderer from scanline renderer to mental ray renderer.

1. Choose the **Render Setup** tool from the **Main Toolbar**; the **Render Setup: Default Scanline Renderer** dialog box is displayed.

2. In this dialog box, click on the **Renderer** drop-down list and select **NVIDIA mental ray**; the **Default Scanline Renderer** is changed to **NVIDIA mental ray** renderer.

3. Choose the **Maximize Viewport Toggle** tool to view all the viewports and activate the Top viewport. Choose **Create > Lights** from the **Command Panel**. By default, the **Photometric** option is displayed in the drop-down list below the **Lights** button.

4. Choose the **Free Light** tool from the **Object Type** rollout; the **Photometric Light Creation** message box is displayed, refer to Figure 12-19. Next, choose **Yes** from this message box.

   Also, various rollouts are displayed in the **Command Panel**.

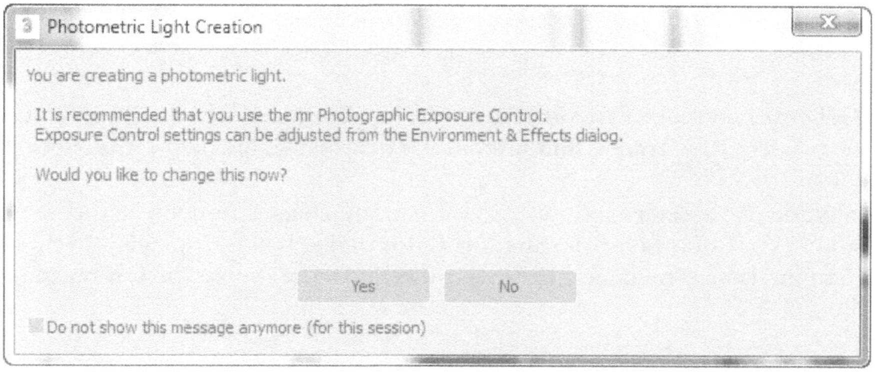

Figure 12-19 *The* ***Photometric Light Creation*** *message box*

5. Click at the center of the room; the free light is created with the name *PhotometricLight001*. Enter **rooflight1** in the **Name and Color** rollout.

6. Choose the **Select and Move** tool and align *rooflight1* in the viewports, as shown in Figure 12-20.

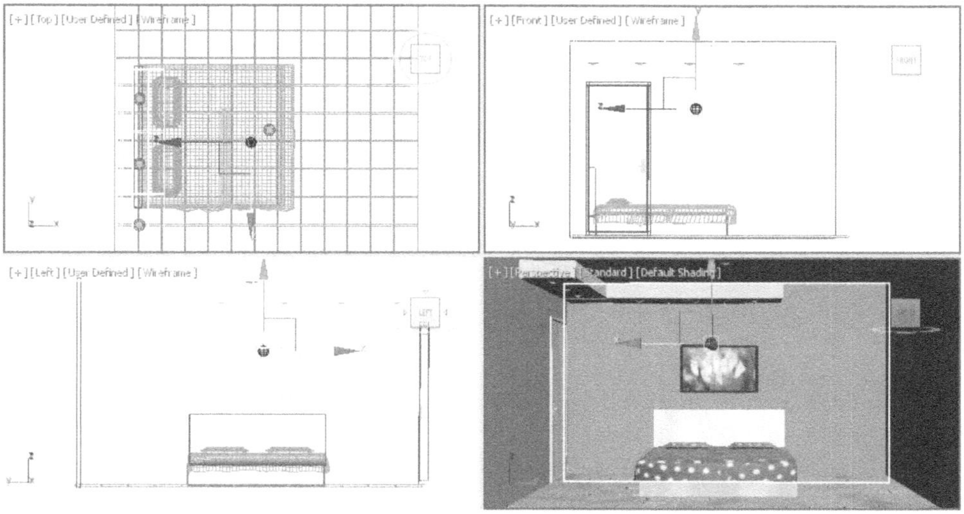

Figure 12-20 *The rooflight1 aligned in the viewports*

Next, you will modify the shape of *rooflight1*.

7. Choose the **Modify** tab in the **Command Panel.** Next, in the **Templates** area, select **40W Bulb** from the drop-down list.

8. In the **Emit light from (Shape)** area of the **Shape/Area Shadows** rollout, make sure the **Point** option is selected in the drop-down list.

Next, you will modify the parameters of *rooflight1* to lit the room with proper intensity, colors, and shadows.

9. In the **General Parameters** rollout, make sure the **On** check box is selected in the **Shadows** area. Next, select **Ray Traced Shadows** in the drop-down list below this check box.

10. In the **Shadow Parameters** rollout, enter **0.7** in the **Dens.** spinner. Next, choose the **Color** color swatch; the **Color Selector: Shadow Color** dialog box is displayed. In this dialog box, enter **15** in the **Red**, **Green**, and **Blue** spinners and then choose the **OK** button to close the dialog box.

11. Make sure the **Uniform Spherical** option is selected in the drop-down list located in the **Light Distribution (Type)** area.

    The **Uniform Spherical** option is used to spread light in all directions.

12. Make sure the **Kelvin** radio button is selected.

13. Set the value **6500** in the **Kelvin** spinner; the color swatch located next to the **Kelvin** spinner is changed to white. Next, choose the **Filter Color** color swatch located below it; the **Color Selector: Filter Color** dialog box is displayed. In this dialog box, make sure the value in the **Red**, **Green**, and **Blue** spinners is set to **255** and then choose the **OK** button to close the dialog box.

Note

The value in the ***Kelvin*** *spinner and the color in the* ***Filter Color*** *color swatch together define the color shade of the light being created. You can use different combinations of the spinner value and color shade in the color swatch to achieve a variety of light color shades.*

Next, you will set the intensity of *rooflight1*.

14. In the **Intensity** area of the **Intensity/Color/Attenuation** rollout, select the **lm** radio button; the value in the first spinner located below the **lm** radio button changes and displays the intensity of the light in Lumen. Next, set the value **40** in this spinner; the intensity of the light is changed.

15. Choose **Rendering > Environment** from the menu bar; the **Environment and Effects** dialog box is displayed. In this dialog box, Make sure **mr Photographic Exposure Control** is selected in the drop-down list of the **Exposure Control** rollout. Next, click on the **Render Preview** button to see the render preview in the window available in this rollout.

16. In the **mr Photographic Exposure Control** area, select **Physically Based Lighting, Indoor Nighttime** in the **Preset** drop-down list. Next, select the **Exposure Value (EV)** radio button, if not already selected and then enter **0.6** in the **Exposure Value (EV)** spinner; the preview of the render is updated in the **Exposure Control** rollout.

17. In the **Image Control** area, set the values as given next:

    Highlights (Burn): **0.5** Midtones: **1.25** Shadows: **0.1**

    Now, you will render the scene to view the effect of modifying *rooflight1* parameters in the scene.

18. Activate the perspective viewport. Next, choose the **Render Production** tool from the **Main Toolbar**; the rendered image is displayed, refer to Figure 12-21.

Figure 12-21 *The rendered image*

Creating Wall Lights

In this section, you will create wall lights by using the **Target Light** tool. You will also use IES file to achieve a specific light pattern.

1. Switch from the Left viewport to the Right viewport. Now, activate the Front viewport. Choose **Create > Lights** from the **Command Panel**. Next, choose the **Target Light** tool from the **Object Type** rollout; various rollouts are displayed in the **Command Panel**.

2. Click at a point and drag the cursor downward. Next, release the left mouse button; a target light is created. In the **Name and Color** rollout, enter **spot light1**. Next, choose the **Select and Move** tool and then select *spot light1* and *spot light1.Target*. Now, align them in the viewports, as shown in Figure 12-22.

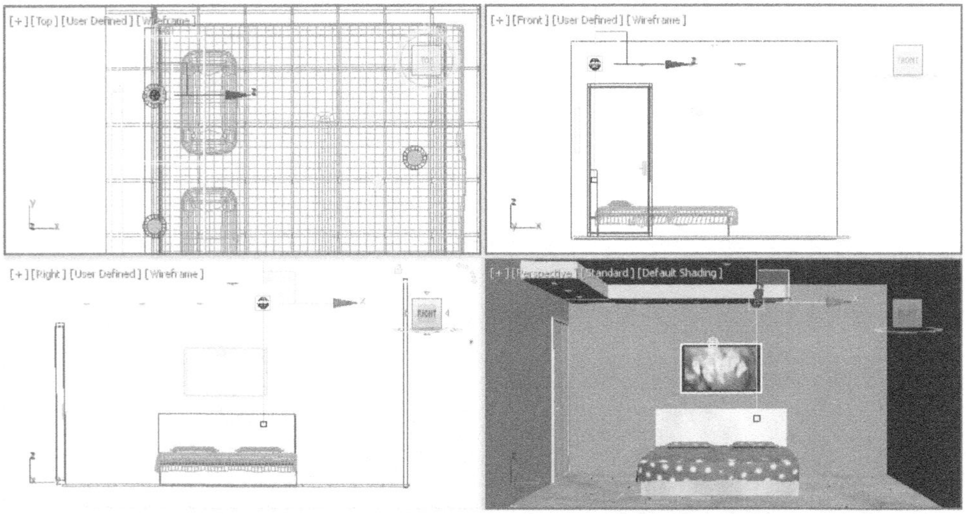

Figure 12-22 *The spot light1 and spot light1.Target aligned in the viewports*

Next, you need to modify the parameters of *spot light1*.

3. Select *spot light1* and then choose the **Modify** tab in the **Command Panel**. In the **General Parameters** rollout, select the **Photometric Web** option in the drop-down list from the **Light Distribution (Type)** area; a **Distribution (Photometric Web)** rollout is displayed in the **Command Panel**.

   Now, you will use a standard photometric IES file to create a specific lighting pattern.

   These IES files are standard files provided by light manufacturers and are available on the internet for free download.

4. In the **Emit light from (Shape)** area of the **Shape/Area Shadows** rollout, make sure the **Point** option is selected in the drop-down list.

5. Choose the **< Choose Photometric File >** button from the **Distribution (Photometric Web)** rollout; the **Open a Photometric Web File** dialog box is displayed. As the project folder is already set, the *photometric* folder of this project is displayed in the **Look in** drop-down list. Next, select the **1.ies** file from this dialog box and then choose the **Open** button; the label of the button is replaced by the name of the selected file. Also, the lighting pattern is displayed in the window located above this button. Next, set the value in the **Y Rotation** spinner to rotate *spot light1*, as shown in Figure 12-23.

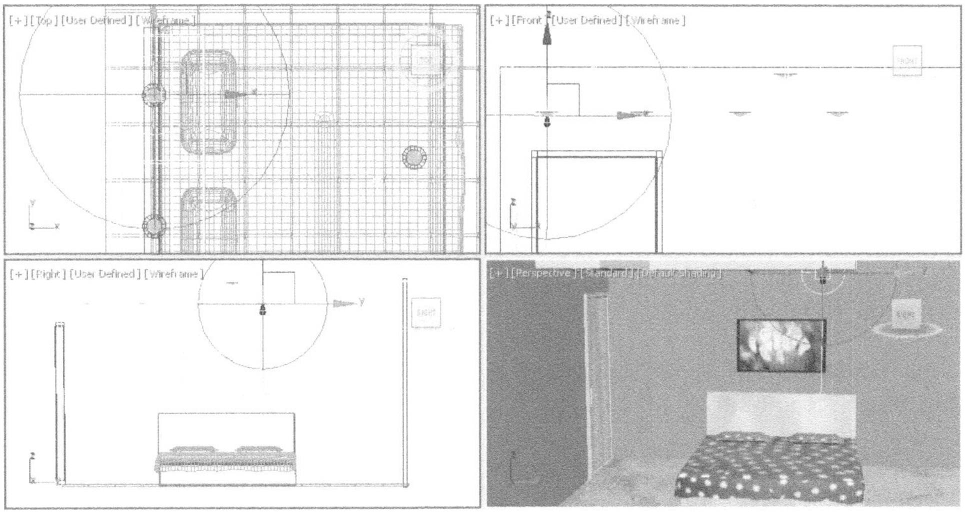

Figure 12-23 The spot light1 rotated

Next, you need to set the intensity and color of *spot light1*.

6. Select the radio button located near the **Kelvin** Spinner; the **Kelvin** spinner is activated.

7. Choose the **Filter Color** color swatch located below it; the **Color Selector: Filter Color** dialog box is displayed. In this dialog box, set the value in the **Red**, **Green**, and **Blue** spinners as given next:

   Red: **164** Green: **165** Blue: **112**

   Choose the **OK** button to close the dialog box; the color in the **Filter Color** color swatch is changed.

8. In the **Intensity** area of the **Intensity/Color/Attenuation** rollout, select the **lm** radio button; the value in the first spinner located below this radio button changes and displays the intensity of the light in Lumen. Next, set the value **4** in this spinner.

9. In the **Far Attenuation** area of the **Intensity/Color/Attenuation** rollout, select the **Use** and **Show** check boxes. Next, set **10** and **34** in the **Start** and **End** spinners, respectively.

10. Activate the Perspective viewport. Next, choose the **Render Production** tool from the **Main Toolbar** to view the render.

   Next, you will add glow to the source of *spot light1*.

11. Choose **Rendering > Material Editor > Compact Material Editor** from the menu bar;

the **Material Editor** dialog box is displayed. In this dialog box, choose **Utilities > Reset Material Editor Slots** from the Material Editor menu bar; all slots in the **Material Editor** dialog box are empty. Now, select any empty slot in it and modify its name in the **Material Name** text box to *lamp dome material*.

12. Select the **Color** check box in the **Self-Illumination** area of the **Blinn Basic Parameters** rollout. Next, choose the **Color** color swatch and change its color to white.

13. Choose the **Diffuse** and **Specular** color swatches from the **Blinn Basic Parameters** rollout and change its color to white.

14. Set the value **81** in the **Specular Level** spinner and **51** in the **Glossiness** spinner of the **Blinn Basic Parameters** rollout.

15. Select *lamp holder dome* from the Scene Explorer. Next, choose the **Assign Material to Selection** button from the **Material Editor** dialog box; the self illuminating material is applied to *lamp holder dome*.

16. Choose the **Render Production** tool from the **Main Toolbar**; the rendered image is displayed, as shown in Figure 12-24. You will notice glow at the source of *spot light1*.

Figure 12-24 *The rendered image*

17. Activate the Right viewport and select *spot light1* and *spot light1.Target*. Next, press and hold the SHIFT key and drag the cursor to the adjacent lamp holder located at the left of *spot light1*. Release the left mouse button; the **Clone Options** dialog box is displayed. In this dialog box, select the **Copy** radio button. Set the value **2** in the **Number of Copies** spinner and choose the **OK** button; two copies of *spot light1* and *spot light1.Target* are created in the viewports. Next, choose the **Select and Move** tool and align these lights in the viewports, as shown in Figure 12-20.

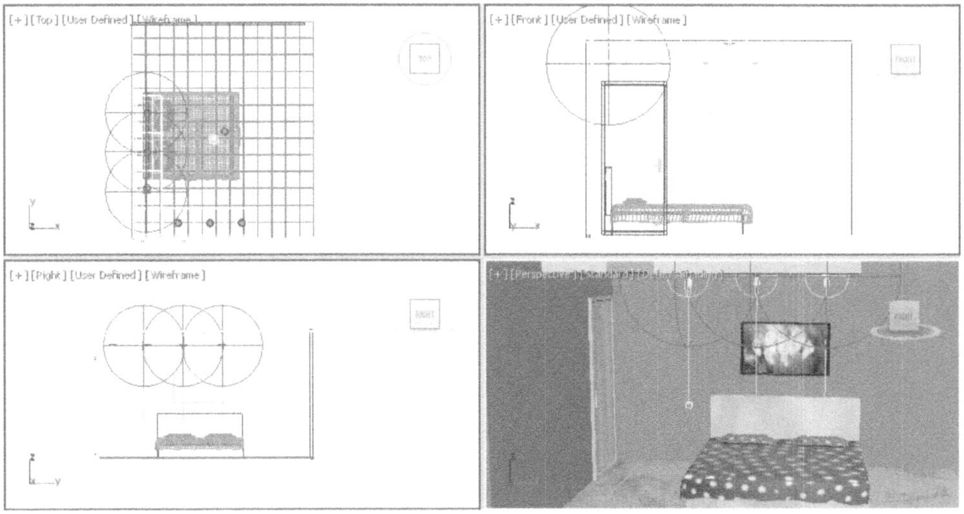

Figure 12-25 *Copies of spot light1 and spot light1.Target aligned in the viewports*

18. Assign the *lamp dome material* to *lamp holder dome 001, lamp holder dome 002* and *lamp holder dome 006* to add glow at the source of lights in the scene.

Note

You can create copies of spot light1 and spot light1.Target and place them at the remaining spot lights of the false ceiling. You need to assign the lamp dome material to remaining lamp holder domes.

Saving and Rendering the Scene

In this section, you will save the scene and then render it. You can also view the final rendered image of this model by downloading the file *c12_3dsmax_2017_rndr.zip* from *www.cadsofttech.com*. The path of the file is as follows: *Textbooks > Animation and Visual Effects > 3ds Max > Autodesk 3ds Max 2017 for Novices.*

1. Choose **Save** from the **Application** menu.

2. Activate the Perspective viewport. Next, choose the **Render Production** tool from the **Main Toolbar**; the rendered image is displayed, refer to Figure 12-26.

   You can improve the rendered output further by performing the following steps:

3. Drag the rendered frame window upward to see the parameters available below it. Next, move the slider below the **Final Gather Precision** text box toward right so that **Low** is displayed in that text box.

4. In the **Trace/Bounces Limits** area, enter **2** in the **FG Bounces** spinner. Next, render the output. You will notice the change in the rendered output.

Figure 12-26 *The rendered image*

EXERCISES

The rendered output of the models used in the following exercises can be accessed by downloading the *c12_3dsmax_2017_exr.zip* file from*www.cadsofttech.com*.The path of the file is as follows: *Textbooks > Animation and Visual Effects > 3ds Max > Autodesk 3ds Max 2017 for Novices.*

Exercise 1

Create a scene using the standard primitives and target spot lights, as shown in Figure 12-27.

(Expected time: 15 min)

Exercise 2

Create a scene using the standard primitives and target spot lights, as shown in Figure 12-28.

(Expected time: 30 min)

Figure 12-27 *The target spot lights illuminating the wall*

Figure 12-28 *Various lights illuminating the room*

Hints:

1. Create the hollow effect on the roof using the **ProBoolean** compound object and use the **Free Point** photometric light.

2. Create a **Target Spot** standard light for highlighting the object. You need to add the **Volume** effect using the **Atmospheres & Effects** rollout for the projector effect. Also, apply the projector map using the **Advanced Effects** rollout to get a realistic effect.

Index

Made in United States
Cleveland, OH
05 November 2026

47673680R00140